The Tao of Measurement:
A Philosophical View of Flow and Sensors

The Tao of Measurement:
A Philosophical View of Flow and Sensors

By Jesse Yoder and Dick Morley

Setting the Standard for Automation™

67 T.W. Alexander Drive
P.O. Box 12277
Research Triangle Park, NC 27709

10 9 8 7 6 5 4 3 2 1

ISBN: 978-0-876640-91-3

Notice
The information presented in this publication is for the general education of the reader. Because neither the author nor the publisher has any control over the use of the information by the reader, both the author and the publisher disclaim any and all liability of any kind arising out of such use. The reader is expected to exercise sound professional judgment in using any of the information presented in a particular application. Additionally, neither the author nor the publisher have investigated or considered the affect of any patents on the ability of the reader to use any of the information in a particular application. The reader is responsible for reviewing any possible patents that may effect any particular use of the information presented.

Any references to commercial products in the work are cited as examples only. Neither the author nor the publisher endorses any referenced commercial product. Any trademarks or tradenames referenced belong to the respective owner of the mark or name. Neither the author nor the publisher makes any representation regarding the availability of any referenced commercial product at any time. The manufacturer's instructions on use of any commercial product must be followed at all times, even if in conflict with the information in this publication.

Library of Congress Cataloging-in-Publication in process

This book is dedicated with thanks to my mother and her prayers,
and to the loving memory of my father.
—Jesse Yoder

To Jesse, and to the silk purse staff, Susan Colwell and Deb Morrison.
Tuff going and they managed us well.
—Dick Morley

Acknowledgments

I developed the ideas for this book over a period of 35 years. Some of my knowledge of mathematics and set theory go back to tutorials on set theory at The Rockefeller University with Leslie Tharp. I discussed my ideas for Circular Geometry at length on the Geometry Forum with John Conway and others. Some of this discussion is reproduced at www.circulargeometry.net.

I am grateful to my professors at The Rockefeller University for their patient critiques of my papers written for weekly tutorials in philosophy of mind and other areas. Chief among these were Donald Davidson and Joel Feinberg. Norton Batkin was a good friend and colleague during this time. At the University of Massachusetts at Amherst, Gareth Matthews shepherded me through the dissertation process with great skill, and was also a friend to me.

Belinda Burum served as my editor for the book, and is responsible for any catchy phrases that may appear in the book. She has been a wonderful friend to me for 25 years, served as a partner in building Flow Research, and has been a source of inspiration throughout this time. My assistant, Nicole Riordan, has brought both humor and organization to my work life, and has been a joy to work with. For more than 10 years, I have worked with Matt Migliore of Flow Control, who has helped in topic selection and then in publishing many of my articles.

Dick Morley helped me develop my ideas through multiple discussions in order to make this book possible. I would like to thank him for his incisive contributions to the book in the form of "Morley's Points." I am grateful to Susan Colwell of ISA for guiding us both through the process of completing the book.

Half of this book is about temperature, pressure, and flow. Many of the ideas expressed here were developed in the course of 24 years of market research, most of it at Flow Research. Flow Research would not be possible without the support of our many clients over many years. I cannot thank them enough for their support. I would single out four as especially important and helpful to me: Mark Heindselman of Emerson Process Management, Mike Touzin of Endress+Hauser, Matt Olin of Sierra Instruments, and Randy Brown of Fluid Components Int'l. They have been both colleagues and friends.

More than anyone else, I am grateful to Vicki Tuck for 23 years of love, joy, and humor. More than anyone else, she has been my muse and my reason for being. Philosophy and flow remain my greatest loves, apart from Vicki.

—Jesse Yoder

Table of Contents

About the Authors

Jesse Yoder, Ph.D. (www.flowresearch.com), is president of Flow Research, Inc., a company he founded in 1998, which is located in Wakefield, MA. He has 28 years of experience as an analyst and writer in process control. He has authored more than 180 market research studies in industrial automation and process control and has written more than 230 published journal articles on instrumentation topics. He has published articles in *Flow Control, Processing, Pipeline & Gas Journal, InTech* magazine, *Control*, and other instrumentation publications. Study topics include Coriolis, magnetic, ultrasonic, vortex, thermal, differential pressure, positive displacement, and turbine flowmeters. He has authored two separate six-volume series of studies on gas flow and oil flow. Dr. Yoder is a regular speaker at flowmeter conferences, both in the U.S. and abroad.

Dr. Yoder studied philosophy at the University of Maryland, The Rockefeller University, and the University of Massachusetts Amherst, where he received his Ph.D. in 1984. He served as an adjunct professor of philosophy for ten years at the University of Massachusetts Lowell and Lafayette College. In 1989 he co-founded the InterChange Technical Writing Conference, which he directed for six years. He has become a world-renowned authority and expert in the area of flow measurement and market research. As an entrepreneur, author, consultant, and inventor, he has helped define the concepts used in flow measurement, and is widely respected as an innovator in this field.

Yoder lives in Wakefield, MA where he enjoys racquetball, bird-watching, and reading detective novels.

Richard E. Morley (Morley@alum.mit.edu), best known as the father of the programmable logic controller (PLC), is a leading visionary in the field of advanced technological developments. An entrepreneur whose consistent success in the founding of high technology companies has been proven through more than three decades of revolutionary achievement, Morley has — among his many accomplishments — more than 20 U.S. and foreign patents, including the parallel interface machine, hand-held terminal, and magnetic thin film. His MIT-based background in physics has provided insight into becoming an internationally recognized pioneer in the areas of computer design, artificial intelligence, automation, and futurism. As an inventor, author, consultant and engineer, Morley has provided the R&D community with world-changing innovations. His peers have acknowledged his contributions with numerous awards, honors, and citations. Morley's medals of achievements are from such diverse groups as Inc. magazine, the Franklin Institute, the Society of Manufacturing Engineers, and the Engineering Society of Detroit. He has also been inducted into the Manufacturing Hall of Fame.

Because the world is round, it turns me on.
– from "Because" by the Beatles

Chapter One

Beginning Point

The idea for this book came about in 2003. At that time, Dick Morley was hosting his annual "Geek Pride Day." This was a highlight of June, when he invited 60 or so intellectual types and business leaders for a day of discussion and idea exchanges. The event was hosted on his farm in Mason, New Hampshire. The food was always good, and the presentations were both interesting and informative. While most people came from the local area, some came from as far away as Pennsylvania for this event.

At that time, I conceived of the idea of writing a book that talked about fluid flow, instrumentation and measurement. Measuring fluid flow is part of a broader range of tasks involving sensing and measurement. I figured that with my background in philosophy and flow, and Dick's knowledge of physics, we could analyze these topics from different perspectives. After discussing this idea back and forth for quite a few years, we approached ISA with the idea of publishing the book. This is the book that resulted from our efforts.

We selected the title *The Tao of Measurement: A Philosophical View of Flow and Sensors* because "Tao" conveys the idea of "underlying principle" or "path." Although this book is not based on Taoism or the teachings of Lao-Tzu, it does attempt to reach beyond a purely scientific discussion of the topic in a meaningful and useful way.

The structure of this book is as follows: Chapter 1 - Beginning Point, followed by Chapter 2 - Temperature; Chapter 3 - Pressure; Chapter 4 - Flow; Chapter 5 - Time; Chapter 6 - Length; Chapter 7 – Area; and Chapter 8 - Theory of Measurement. Dick Morley shares his perspective at the end of every chapter, including this one. The book concludes with Chapter 9, written by Dick.

This chapter is written more like an executive summary than an introduction. As you read through this chapter, you will gain an insight into the main points made in the book. You can then go to the chapter or chapters that interest you most. For example, if your area of interest is fluid flow, you may want to start with Chapter Four. If you are fascinated by time, go to Chapter Five. Each chapter is designed as a stand-alone treatment of the topic of that chapter. The only exception is Chapter Eight, which attempts to bring the areas of sensing and measuring together into a broader theory. Of course, we want you to read the entire book, but that's up to you.

Scope

This book covers both sensing and measuring. These are broad topics, but we have selected six subjects within those topics that are of special interest. First, we focus on sensing and then measuring these three parameters:

- Temperature
- Pressure
- Flow

The discussion of sensing temperature, pressure and flow centers mainly on the different types of sensors that are used to sense these parameters. We then talk about how they are measured. There are a number of different technologies involved in sensing and measuring temperature, pressure and flow, and they are discussed in Chapters 2, 3 and 4.

In measuring, we have chosen three parameters that are especially interesting and important:

- Time
- Length
- Area

Chapters 5, 6 and 7 center on the units used to measure these parameters. The structure of each of these chapters is similar. The first part of each chapter explores the historical origins of many of the units that we still commonly use to measure time, length and area. The origins of the terms and practices in use today go back over two thousand years, to the Greeks and Romans in some cases. The second part of each of these three chapters describes the generally accepted methods for measuring the parameter in question. The last part of each chapter proposes a new perspective that hopefully avoids some of the pitfalls in the way measurement is done today. These new ideas include flowtime, Wide Line Geometry and Circular Geometry. The last section of each chapter includes a list of the units of measurement that are associated with the subject of the chapter.

Chapter Two: Temperature

Chapter Two begins by tracing the origins of the thermometer. Galileo is credited with inventing the first thermometer in the 1590s. It was an air thermometer made of a glass bulb with a long tube attached. While Galileo called it a *thermoscope*, it worked enough like a thermometer to earn him the title of the inventor of the thermometer.

There are three main temperature scales that are commonly used today:

- Fahrenheit
- Celsius
- Kelvin

This chapter compares the origins of these scales and shows how they compare to each other. Another important scale is the Rankine scale, which is mainly used for specialized engineering applications.

Most of the rest of Chapter Two is devoted to a discussion of the following five types of temperature sensors:

- Thermocouples
- Resistance Temperature Detectors (RTDs)
- Thermistors
- Infrared Thermometers
- Fiber Optic Sensors

The first three types on this list are contact temperature sensors, while infrared and fiber optic are non-contact types. Infrared thermometers sense the temperature and also display it, so they are also measuring devices.

There are a number of different types of *thermocouples* that are distinguished according to the type of metals making them up. Each one is designated with an alphabetical letter, and they are designed for different temperature ranges. The two main types of RTDs are wirewound and thin film. *Thermistors* are less robust than RTDs and have a more narrow temperature sensing range, but are more sensitive than RTDs within that range. They are widely used for food applications. *Infrared thermometers* are used to measure temperature at a distance. Many of them are used in a "point and shoot" method of measuring, where the temperature is read off the material that the infrared beam is aimed at. *Fiber optic* sensors use optical fibers in making temperature measurements.

This chapter concludes with a discussion of the relative advantages and disadvantages of the different types of temperature sensors. For example, thermocouples can be used at higher temperatures than RTDs and respond more quickly, but RTDs are more stable than thermocouples and are more accurate than thermocouples within their more limited temperature range.

Chapter Three: Pressure

This chapter begins by looking at different definitions of pressure. Most of the chapter is then devoted to pressure transmitters, especially differential pressure transmitters, which play an important role in flowrate measurement. Transmitters accept input from a sensor, amplify it, and convert its value into a signal that can be transmitted to a recording or controlling device.

Differential pressure (DP) transmitters are one of four types of pressure transmitters. These four types are as follows:

- Multivariable
- Differential pressure
- Gage
- Absolute

The early part of the chapter traces the history of differential pressure measurement. Bernoulli developed his famous equation for flowrate calculation in 1738. Differential pressure transmitters rely on a constriction in the flowstream to create a difference between upstream and downstream pressure. Differential pressure transmitters then use this difference in pressure as a variable in Bernoulli's equation to calculate the rate of fluid flow through a pipe.

The constriction is created by what is called a *primary element.* The most widely used type of primary element is the orifice plate. The first commercial orifice plate was introduced in 1909. Since that time a great deal of work has been done by the American Gas Association (AGA) and the American Society of Mechanical Engineers (ASME) in studying DP flowmeters and how they perform under various conditions. Both the AGA and ASME have issued standards for the use of DP flowmeters as a result of their research.

Chapter Three also describes the following three pressure sensor technologies:

- Piezoresistive
- Strain gage
- Capacitive

These three technologies are used by differential pressure sensors to detect upstream and downstream pressure in differential pressure transmitters.

Differential pressure sensors are an essential component of DP flow measurement, but primary elements are equally important, since they create the difference in pressure in the flowstream that makes DP flow measurement possible. The main kinds of primary elements are:

- Orifice plates
- Venturi tubes
- Pitot tubes
- Flow nozzles
- Wedge elements

Orifice plates consist of a round, usually metal plate with an opening (orifice) in it to enable fluid to pass through. They are used to measure the flowrate of liquid, steam and gas flowstreams. *Venturi tubes* are especially useful for fluids with suspended solids because of their wide openings. *Flow nozzles* are especially suited to measuring high-speed flows, and can handle fluids containing particles. *Pitot tubes* are widely used for measuring air flow. Averaging Pitot tubes contain more than one measuring port and provide higher accuracy than single port Pitot tubes. They are also used to measure fluid flow in large pipes. *Wedge elements* are manufactured by only a few companies. They are designed to handle fluids with a high solids content.

Chapter Four: Flow

Chapter Four is devoted to discussing the different types of flowmeters used to measure flowrate, which is often shortened to "flow" when there is little chance of misunderstanding. It begins by distinguishing between new-technology flowmeters and traditional technology flowmeters. New-technology flowmeters were introduced after 1950 and are more the object of product development today than traditional technology flowmeters. Traditional technology flowmeters were introduced before 1950 and their performance is typically not at the same level as that of new-technology flowmeters.

The following is a list of the different types of flowmeters:

New-Technology Flowmeters

- Coriolis
- Magnetic
- Ultrasonic
- Vortex
- Thermal

Traditional Technology Flowmeters

- Differential Pressure
- Positive Displacement
- Turbine
- Open Channel
- Variable Area

Each section discussing these flowmeters includes:

- Principle of Operation

- Paradigm Case Application
- Applications

The Paradigm Case Application section includes a specification of the conditions that are ideal for the best operation of this type of flowmeter. For example, the paradigm case application for turbine flowmeters is for measuring the flowrate of **clean gases or clean low-viscosity liquids flowing at medium to high speeds**. Some sections also discuss the different types of the meter in question. For example, there are eight main types of turbine flowmeters.

The chapter concludes with a discussion of the emerging technologies of sonar and optical flowmeters. These two technologies were introduced after the year 2000 and are currently manufactured by only a handful of suppliers. *Sonar* flowmeters are used in water & wastewater and pulp & paper applications, among others. One major use for *optical* flowmeters is emissions monitoring.

Another technology not discussed in Chapter Four that can also be considered an emerging technology is multiphase flow measurement. Multiphase flowmeters measure the mixture of oil, gas and water that comes out of an oil well. Research on these began in Norway shortly after 1980. The motivation for the research was the declining performance of the North Sea oil fields. Multiphase flowmeters perform a valuable measurement, but they are very expensive. The technology is still in its early stages, and accuracy is a major concern for these flowmeters.

Chapter Five: Time

Time and time-keeping is a topic that almost everyone is intimately familiar with. Our society today is increasingly "schedule-driven." Of course, in order to have social events such as school classes, football games and business hours of operation, it is necessary to have a common method of measuring time that is available to all and is generally agreed upon. For the familiar daily scheduling of our activities, time-keeping today is provided by clocks and watches.

While there is fairly wide consistency in the units of time used for clocks and watches, beyond that things are not as simple as they may seem. There are 24 different time zones around the world, most of them one hour apart. Because of the earth's rotation, the sun rises and sets at different times in different places. Time zones take this into account so that people in different parts of the world can sleep through the night and enjoy the sunrise, if they are up early enough. It also better serves the needs of business and industry. Anyone who does business in other countries, or indeed between the far ends of the larger countries, needs to take time differences into account. For example, when it is 11:00 am on the East Coast of the United States, it is already 5:00 pm in much of Europe, so most business-oriented phone calls to Europe from the United States (for example) should be made before then.

Our bodies become adjusted to the time zone we are in. As anyone who has traveled through three or more time zones by airplane knows, making the adjustment to a new time zone can be difficult. If you fly from New York to Los Angeles, for example, the clock may say 8:00 pm there but for you it is 11:00 pm. I was able to traverse all 24 time zones in one trip to Australia by flying from Boston to New York and then east to Dubai to Perth, Australia. I then continued east to Sydney, Australia, to Dallas, and back to Boston. (This took considerably less than the 80 days it took in Jules Verne's book, "Around the World in 80 Days.") Modern day transportation has made it possible to travel to even remote parts of the globe in considerably less than 24 hours.

Chapter Five describes early methods of time-keeping, which mainly related to agricultural needs. Much of the life of the ancient Egyptians centered around the Nile River, which flooded from the end of June until October. Crops were planted during this time, and grew until late February. Crops were harvested from then until the end of June. The Egyptians measured the flood level of the Nile with a stick, which became the basis for a "Nile Year." They developed a workable 12 month calendar that allotted 30 days to each month and added five days at the end of the year.

The Romans used sundials and water clocks to measure time. In early Greece, water clocks were used to measure the length of time allowed for pleading in court. Orators would not infrequently ask for the water flow to be stopped to allow them more time to talk. In the 8th century A.D., the hourglass was invented. Hourglasses were used to measure specific periods of time, and they had the advantage that they worked both day and night. By the 15th century they were being used in England to measure the length of sermons.

Our division of the day into two 12 hour periods seems to go back to early sundials, many of which were divided into 12 equal parts. Our division of hours into 60 minutes and of minutes into 60 seconds is rooted in the Babylonian number system, which was based on the number 60.

The Invention of Mechanical Clocks

Christiaan Huygens is credited with inventing the pendulum clock in 1656. By the mid-1600s, pendulum clock movements were in common use, making it possible to add a minute hand and a second hand to the early mechanical clocks.

In the 18th century, clocks were still expensive. Watches began to be made in the late 1600s, and by the early 1800s, clockmakers had begun to work on methods of mass production, which made both clocks and watches accessible to many more people. In the late 1920s an engineer at Bell Laboratories discovered that using a quartz crystal greatly increased the accuracy of clocks and watches. This invention was followed by the development of atomic clocks, which provided even greater accuracy.

In the 1960s, digital clocks began to be used, and digital watches were first available in the early 1970s.

There is much to be said for the accuracy of today's clocks and watches. Clocks are everywhere, including on televisions and radios, and it is difficult not to be aware of the time. Radio and television programming is scheduled to begin and end at precise times, as are most sporting events, public meetings and workdays. Planes and trains are scheduled to depart to the exact minute, although they are notorious for sometimes "running late." All this precision and accuracy have facilitated the operation of societies composed of millions of people, and in some cases, this extensive scheduling is required for safety reasons. If pilots were to fly planes without adhering to schedules, there would soon be chaos in the skies.

Clock Time and Biological Rhythms

Having granted that there are many advantages to "living by the clock," there is something vaguely disquieting about having one's life so completely governed by what is a mechanical and somewhat arbitrary division of time into 24 hours, 60 minutes and

60 seconds. For many people, clocks dictate when we rise in the morning, when we have lunch, when we have dinner and when we go to bed. However, human beings are biological beings, not mechanical devices, and our needs do not necessarily conform to the requirements of a 1440 minute day. The units of time that we are governed by are based on our ability to divide time into many tiny equal segments; they are not designed to coincide with the flow of our biological rhythms.

Decimal Time and Flowtime

Chapter Five also discusses the history and merits of decimal time and "flowtime." There are different forms of decimal time. Some divide the day into 10 hours rather than 24 hours. The hour is often divided into 100 minutes rather than 60, and minutes are typically 100 seconds long. The French instituted such a form of decimal time in 1793. However, people were unable to make the adjustment to the change in time and it lasted only 17 months.

Flowtime is a new form of time that is proposed in Chapter Five. Flowtime keeps the 24 hour clock, but allocates 100 minutes to an hour and 100 seconds to a minute. This system of time would be easier to adjust to than was the French decimal time system with 10 hours per day. In addition, with the advent of digital clocks and watches, displaying flowtime would not be as difficult as it is on a typical analog clock.

The chief advantage of flowtime is that it divides time into smaller units. The National Basketball Association already uses a form of decimal time for the last few minutes of most games. Instead of having the clock simply tick off the remaining seconds, each second is divided into tenths of a second, so instead of showing 2 seconds left on the clock, a play might occur with 1.3 seconds left. Since in some cases the outcome of a game can depend on exactly when a shot was made, this allows both players and referees to understand exactly how much time is left in the game and at exactly what time certain plays were made. Time is also divided into tenths or hundredths of a second at certain Olympic events such as swimming and skiing.

Flowtime also has applications for daily life. It is sometimes said that work is often completed in the time allowed, whether that much time is required or not, so someone who is allowed a half-day to complete a task may take the entire half-day, even if it could be completed in 2½ hours. Somehow the time used to finish the work expands to meet the time allotted for the work.

Flowtime provides more units of time within which to complete a task. While these units of time are smaller, they make it easier to divide a task into smaller parts, so if one is allotted a half-day to complete a task, this translates into 4 hours, as in traditional time, but these 4 hours translate into 400 flowtime minutes instead of the traditional 240. Because there are more units of time to work with, it is easier then to break the job into smaller segments and potentially complete it sooner, rather than simply thinking of it as a half-day job. Flowtime facilitates the setting of smaller subtasks within a larger task, which often helps to complete it more quickly.

Any broadscale transition to flowtime will be difficult until flowtime clocks are available. However, as just noted, a form of flowtime is already used in some sporting events and it is possible to apply flowtime mentally by thinking of hours as having 100 minutes and minutes as having 100 seconds. Digital clocks should make it easier to display flowtime than with analog clocks. Digital clocks display the minutes and seconds, while analog clocks only approximate them. Chapter Five includes a table for converting traditional time to flowtime.

Chapter Six: Length

Length may seem like an unusual choice of topic for this book, but length is one of the most fundamental quantities in our lives. In fact, length is almost as pervasive as time in our daily and business lives. Surrounding and related to length is a cluster of vital units of measurement that we encounter daily. These include distance, miles, inches, meters, points, lines and continuity. The distance between two points is measured in units of length, and it is often important to know how far it is to some location, or how high or wide an object is. These parameters are all interconnected, and discussing length requires a discussion of many of these other parameters as well.

Arriving at a Unit of Measurement

Chapter Six begins with a look at the historical origin of some of the terms we use today to designate units of length. The idea of a foot as consisting of 12 inches goes back to Roman times. Around the 12th century, the idea arose of defining a foot as 1/3 of a yard. King Henry I of England decreed the yard to be the distance from his nose to the tip of his outstretched thumb. Later on, in the 1740s, the British government attempted to establish the length of a yard through the construction of a brass bar of the required length. In 1834, the two existing standard yard bars were destroyed by fire. The effort was renewed and continued until 1855, when a bar made with advanced thermometers to account for thermal expansion was declared the legal standard.

In 1866, the use of the metric system became permissible in the United States, due to an Act of Congress. After this time, the focus turned to defining the length of a standard meter, with the length of a yard being defined in terms of the length of a meter. Eventually, the idea of using a standard bar to define the length of a meter was abandoned, and a meter's length was defined in terms of the path length traveled by light in a vacuum during a time interval that is a tiny fraction of a second.

Much of the discussion in Chapter Six relates to analyzing the relationship between a point and a line. A line is typically drawn between two points, and the distance between two points is often represented as a line. The number line is similar, but it mainly figures in discussions about geometry and the existence of points in the line or on the line. In Euclidean geometry, the number line is often conceived of as a straight line consisting of infinitely many points stretched out to infinity in both directions. It is always possible to put another point between any two points, so there are infinitely many points.

Oh Line, Where Is Thy Point?

A common mathematical conception is that the number line is made up of points, and that another point can lie between any two points. Each point is conceived of as having no dimension or area, so there can be infinitely many points in the number line. The main problem with this idea is that if lines are made up of points with no area, it doesn't matter how many of them you put side by side – the line will still not have area or dimension. Mathematicians try to fix this problem by bringing in infinity, and saying that infinitely many points make up the number line, but zero times any number is zero, and zero times infinity is still zero. Infinity thus becomes a kind of metaphysical glue that salvages one questionable assumption with another, equally questionable assumption.

A more logical conception is that points lie *on* the line and are not *in* the line. If we say this, then even if points have no area, we can still account for the extension of the line by saying it is an unbroken continuum. We can always put another point between two arealess points since there will always be space for still another point that takes up no area.

There are other problems with saying that points have no area, however, for to do so is to invite Zeno's Paradox, which is discussed in detail in Chapter 6. Zeno's Paradox gets its power from conceiving of a person as being at rest at an arealess point. If we conceive of a person as being located at an arealess point, then when he goes halfway, he will be once again at an arealess point. If we continue to multiply by ½, he will in theory never reach his destination since as long as points take up no area; this is an infinite process. Of course, this process of reaching a destination by going halfway each time requires the traveler to stop at each arealess point, which is not what happens with continuous motion.

Wide Line Geometry

Another approach to this topic is to follow Aristotle's definition of a line as a moving point. If we also say that points have area, we can block Zeno's Paradox. This is because there are no longer infinitely many points on the number line. Instead, the number of points on the line is subject to stipulation for a given measurement. If the measurement is made as 1/1000 of an inch, then there are 1,000 points on one inch of the line. If the measurement is made as one-millionth of an inch, then there are one million points on the line, however small a millionth of an inch is. But specifying the location of any object with reference to the points on the line requires that the location be at one of those points. Fractions of a point or partial points are not allowed. This blocks Zeno's Paradox since at some point it will be necessary to specify a location that is a fractional point on the line. This is not allowed because the only legitimate locations on the line have been specified when the number of points on the line has been specified. In order to capture the desired location, it is necessary to specify a number line with greater precision; that is, with a larger number of smaller points on the line.

If we follow Aristotle's definition that a line is the path of a moving point, and that a point has area, then a line has a width equal to the diameter of the point used to create it. This is not to say that mathematics is about actual drawn lines rather than the idealized lines they represent, but saying that lines have width coincides much more closely with the way we think of lines in the real world. The chalk lines that mark the edge of a baseball field and the 10-yard markers that divide the field in American football all have width; they do not stand for Euclid's "breadthless lengths." Another example is the solid and broken lines that divide a highway into two sides.

Defining Continuity

If we consider a number line to be continuous, and that discrete points with specifiable area lie on the line, how do we define the line itself? A line is a continuous entity. It is hard to define continuity without using negative terms like "unbroken," "uninterrupted" and "nonstop." These in effect define continuity in terms of what it is not. A continuous noise is uninterrupted; a continuous game is one that goes on without stopping, and a continuous line is one that is unbroken. This applies to a line of any length; lines are of finite length and are also referred to as line segments. One way to define a line is "an unbroken extension that has the shape of the path that a point with a diameter of the width of the line would make if it were moved from one end of the unbroken extension to the other."

Measuring Locations when Measuring Distance

The idea that points have area is given additional support by the way we measure distance. If we measure the distance from the Eiffel Tower in Paris to the Van Gogh Museum in Amsterdam, we might consider any location at the Eiffel Tower as a starting point. It could be in front of the Eiffel Tower, to the side, or even underneath the Tower. Likewise, arriving at the Van Gogh Museum could be arriving at the parking lot, simply getting onto the grounds, or parking on the road by the side of the museum. Here we treat the beginning point and the ending point as a point with area that is loosely defined by its proximity to the object in question. We don't treat these two points as being dimensionless points.

American vs. Metric Standards

One of the remaining challenges in measuring length and distance is the difference between different measuring systems. The most obvious one is the difference between the American convention of using inches, feet, yards and miles and the metric system that uses millimeters, meters and kilometers. While most of the world uses the metric system, the United States has been slow to adopt it. Still, it is gaining a hold in the United States. One problem with this partial adoption of the metric system is that there is not always an exact conversion from one to the other, so the conversions that are used are often approximations. The same issue arises for measuring circular area in terms of squares, as is discussed in Chapter Seven.

Chapter Seven: Area

Chapter Seven talks about area as measured in circles and squares. Traditional and Euclidean geometry uses the formula $\pi \times r^2$ for the area of a circle, in which r is the circle's radius. Interestingly, though, r^2 is the area of a square that has four sides, each with a length r. So if $r = 2$, then the area of a square with sides of this length is 4 square inches, and the formula is telling us that $\pi \times 4$ square inch squares fit inside the circle. While this is an accepted mathematical truth, it seems counterintuitive in light of the saying "You can't put a square peg inside a round hole."

In Chapter Seven, I propose using a "round inch" instead of a square inch as a unit of measure for circular area. One round inch is equal to $\pi/4$ from a Euclidean point of view. The advantage of this is that it is then possible to give rational values for the area of a circle without relying on the irrational number π. The formula for the area of a circle in round inches becomes $4 \times r^2$, where r is the radius of the circle. I propose to call this Circular Geometry.

If we add Wide Line Geometry to Circular Geometry, it is possible to give rational values for the areas of three-dimensional structures such as pipes. Pipes can be treated as having a "wide line" boundary. The inside diameter can be used for calculating flowrate, while the outside diameter is used to give the size of the pipe including the width of the pipe wall. This idea can be applied to the flow equation:

$$Q = A \times v$$

Here Q is equal to flowrate, A is the area of the pipe in round inches using the inside diameter (ID), and v is flow velocity. This eliminates π, a source of uncertainty and irrationality in the flow equation.

Why is π necessary to begin with? The necessity for π stems from the fact that square area and circular area cannot both be measured using the same unit of measurement. Just as circular area has irrational values when using square area as a unit of measurement, so square and rectangular area will have irrational values when using round inches as a unit of measurement. We either need to have two separate geometries, or be content with irrational numbers such as π in giving circular area. I would at least like to explore the benefits of using Circular Geometry to analyze circular area. I believe it also has important applications in flow measurement.

Chapter Eight: Sensors and Measurement

Chapter Eight attempts to bring the worlds of sensors and measurement together. It begins with a definition of what a sensor is: A sensor senses changes in a physical quality or property, and responds to them in a predictable say. This representation is a reading of the quality or quantity of the physical parameter being sensed. The response must be predictable, or statable according to some rule, to ensure that the sensor is accurate. This chapter discusses mechanical, electronic and biological sensors.

The signals from sensors are often weak. This is why a converter, transducer or transmitter is needed. Such a device accepts input from the sensor, typically amplifies it, and changes it into a representation of the quality or quantity of the variable that is being sensed. If the transducer or transmitter is electronic, this representation will be in electronic form. Transmitters are so-called because they transmit a signal. Transmitters take the electronic representation of the sensor's input and convert it into a signal that can be transmitted to a recording or controlling device. Typical signals include 0–5 millivolts, 4–20 milliamps and various types of digital signals.

This definition of a sensor also helps explain what goes on with the biological sensors that human beings have. Our eyes, for example, accept input in the form of light and convert this in a predictable way in the visual cortex, which is the part of the brain that processes input from the eyes. The image formed in the visual cortex is then transmitted to and processed by the part of the brain associated with seeing and consciousness. The brain acts like a converter or transducer and turns the image formed in the visual cortex into an image that we see through a process of interpretation.

This analysis shows a link between the physical world and our consciousness. When our brains convert visual images into an image that we see, they do so by creating neural patterns that are associated with our conscious perception. This shows how the mind-body problem that Descartes set up can be solved. Descartes defined mind (unextended) and body (extended) as opposites. But when we examine how the brain works, which is somewhat like an electronic transmitter, we can see that certain neural patterns are associated with specific conscious experiences. Mind and body are not opposites, but just complex aspects of each other.

This chapter also defines what a measuring device is. A measuring device makes use of a standard unit of measurement. This unit of measurement is used to determine a quantity that formulates how much of that unit something has. This unit of measurement could be in the form of a scale, like on a yardstick, or it could take the form of a calculated unit like gallons per minute (gpm) as in a flowmeter.

Probably the biggest difference between sensing and measuring is that there is some type of interaction between a sensor and what is sensed – the sensor responds in a predictable way to the quantity or quality that is being sensed. A measuring device, in contrast, does not respond in this way to what is being measured. It simply determines how much size, length or other quantity of a unit of measurement something has. This is done by performing some type of comparison between the unit of measurement and the object being measured.

A meter is an instrument designed to measure the size, length or quantity of something. For example, a flowmeter measures the amount of fluid passing through a closed pipe or an open channel.

The chapter closes with a list of different types of flowmeters used in scientific and process environments.

Morley's Point:
Why Write the Book?

I must admit that, for me, writing is time-consuming and frustrating, but I get a wonderful feeling when I've finished. It's like farming – you have to plant, nurture and harvest your crops, but then you get to sell and eat them.

Jesse Yoder and I have met off and on for many years. We've been in technical meetings together and we've always talked about doing a book, particularly on fluids and fluid mechanics. This is the book. His background is one of a well-educated thinker and he knows all about the details of the control and management of fluid. I put in my two cents as a physics guy.

Sensing and measuring is a big issue in automation. We have to sense what we're doing and we have to measure what we're doing, all to make more and better beer. This book is not on discrete manufacturing automation but on the control of fluids in a process for the control of quality and quantity. (A note here: Sensing and measurement are separate parts of the control of a process. They are not the same.)

Jesse and I (the yin and yang) have tried to connect our knowledge of sensing and measurement and the physics perspective together so that the book is both useful and entertaining. Who would imagine that a book on fluid control would be interesting to read? We hope we have done that.

Physics and Reality

It is, indeed, strange the way the world changes. Right now in the physics world the question being debated is whether the reality we see is the basic physical reality. But, we still have to "make the beer" so we have certain biases and misconceptions about fluid, automation and the process industry. One of my mantras is that the cost of the process equipment is negligible compared to the output of lowest-cost product for consumers. Get the process and the equipment right and you'll be making a high-quality product – reliably – for many years. My farming background says that you should never pick up anything twice. It doesn't add value. Use the math, use technology and engineering, and deliver the goods. Jesse's review of measurement techniques and sensing techniques is the heart of this book. My role is to make a comment at the end of each chapter for better understanding.

We have many biases and misconceptions that make sense to us but are not true. I had to go back to early 1954 to understand some of Jesse's comments in his portion of the book. I spent several hours refreshing dusty parts of my brain, calling up past MIT lectures. Luckily for me, it all came flooding back. One of my favorite misconceptions is, for example, when you have a glass half-full of water and you float some ice in it. Some people are convinced that as the ice melts, the water level rises – but it doesn't change. The misconception that floating, melting ice will raise the water level is not for us to worry about in this book. But it gives you some idea about how wrong we can be in our biases. Our commonsense approach to life served us well when we were cavemen. This is no longer true.

Sensing and Measurement Are Different

I have two dogs. One is a Doberman cross. The other is a West Highland White Terrier. Their sensing characteristics are immensely acute. When someone I know is driving down my driveway outside the barn and they wag their tails, they indicate that it's a friend. These multiple, accurate sensing and response behaviors lead to a conclusion. Every living creature can do this but computers cannot. These loosely coupled sets of sense and response combine to allow dogs, mosquitoes and mother-in-law dinners to show behavior that allows understanding of their environments. Measurement is essentially the quantification of one of these pre-endured senses and must fit into the mathematical analysis that so much of us endure.

There are only four forces in the universe: gravity, electromagnetic, the weak force and the strong force. That's all there are. A trick I use to make sure I always remember this is to think in five dimensions. Most people are unaware that they think in more dimensions than the geometric three: X, Y and Z, the physical coordinates, but we also need to be aware of gravity and time.

Based on the five dimensions of gravity, time and geometry, the moon goes in a straight line. What? If you recognize gravity as a space distortion groove, the moon dissipates no energy in its universe. Thinking in five dimensions helps to understand why the moon is where it is and how it moves.

We will not cover here the concepts of chaos and black swans. A black swan is an event that is totally unforeseen and usually has an explanation that is meaningless. In other words, Stuff Happens. We design for low risk, but all automation and control engineering means is that you can control most of it, but never all of it. You have to be able to absorb the unknown events and assume that they will happen and that they will be unforeseen.

We talk a little in the book about the small world of quantum mechanics. It helps our understanding of fluid mechanics as well. Some new motion sensors and processes use these kinds of principles and make impressive measuring devices. This is well covered in the technical section of the book. As the famous physicist Dr. Richard Feynman once said, "If you think you understand quantum mechanics, you're wrong." I think a lot of that applies to fluid mechanics as well. We know enough to get by, but we do not know everything.

Some of the soft sensing and measurement processes we have to deal with are temperature, flow, humidity, odor, noise, process control, viewing the process, habits of the personnel, communication and the coffee break. (That's not humor, it's real.) Systems are designed and run by people, and people do things that we don't understand. The black swan is always waiting around the corner.

Pressure, Time, Length and Area

From these criteria, we can derive the process of fluid control automation. In this book we cover the process control parameters of pressure, time, area, length and the derivatives of the subsequent control. You'll find a minimum of mathematics in this book because I suspect most of us have forgotten everything we were taught.

Regardless, we hope you enjoy the book and find that it answers most of your questions about the control and management of fluids. In it we list and describe many sensing mechanisms, reference physics, available sensors and automation processes. Enjoy.

"It doesn't make a difference what temperature a room is, it's always room temperature." – Steven Wright

Chapter Two

The Hot and Cold of Industrial Temperature Measurement

Temperature is vitally important both in industrial environments and in our daily lives. Our physical comfort depends on the air temperature. Both the heating and the air conditioning industries are built on the idea of bringing the air temperature to a comfortable level. Cooking and freezing food involve changing the temperature of food to either make it edible and appetizing, or to preserve it so it can be eaten later. The outside temperature governs the choice of clothes we wear, and millions of people are attracted every year to the warmer climes for vacation and relaxation purposes.

Temperature is also important in industrial process environments. In chemical and food processing plants, temperature measurement plays an important role in maintaining food safety and in keeping chemicals at a safe temperature. Also, certain chemical reactions only take place at certain temperatures. In flow measurement, temperature is an essential measurement in computing mass flow when volumetric flow and pressure are also known.

Just as temperature itself is important, so is measuring temperature. Thermostats have a setpoint that represents the desired temperature for a room or building. The first step in controlling the room temperature at the desired setpoint is measuring it. Measuring the internal temperature of food tells us when it is ready to eat. Our decisions about what to do on a given day or weekend are based on the forecast temperature, along with other weather conditions. In an era when global warming is a recognized fact, scientists track the temperature of the oceans to determine how quickly the earth is warming.

While temperature is among the most measured of physical properties, defining it is not so easy.

The temperature that we experience is a subjective quality of the objects we feel as hot or cold. It is this quality that causes us to experience the sensation of heat or coldness. If the object is "hot," it is at high temperature, while if it is "cold," it is at low temperature. What we sense as "hot" or "cold" is caused by molecular motion, and temperature is a measure of the average kinetic energy of the molecules of a substance. The term *kinetic* means "having to do with motion," so another way of saying this is:

> Temperature is a measure of the average energy of a substance due to the motion of the molecules in the substance. As the average motion of the molecules increases, so does the temperature, and as the average motion of the molecules decreases, the temperature does also.

Change in temperature is also objective, and relates to how two bodies at different temperatures interact with each other. When two bodies are in thermal contact with each other, a form of energy called *heat* flows from the warmer body to the cooler body. The warmer body becomes cooler, and the cooler body becomes warmer. This transfer of energy continues until the temperatures of the two bodies become equal. Once the temperatures of the two bodies become equal, the transfer of energy in the form of heat ceases. Bodies in such a state are said to be in a state of thermal equilibrium.

The Historical Question: How to Measure Temperature

If temperature is understood to be a measure of average molecular motion, how should it be measured? Rather than trying to measure the amount of molecular motion directly, temperature is measured indirectly. By making use of a property that changes in a uniform manner with temperature, temperature can be determined by associating different temperatures with different states of that property.

The instrument used to measure temperature is called a *thermometer*. The term *thermometer* is derived from the Greek words meaning "heat measure." The first recorded use of the word *thermometer* in English occurred in 1633. It is described as "an instrument to measure the degrees of heat and cold in the air."

The liquid-in-glass thermometer is one of the most common means of measuring temperature. This type of thermometer makes use of the fact that liquids generally expand as they are heated and contract when they are cooled. As the temperature rises, the liquid in the glass tube expands, extending along a scale, and the scale on the thermometer indicates the temperature. The temperature is proportional to the length of the liquid column. Although it is no longer commonly used because of its toxicity, mercury was once favored because of its high coefficient of thermal expansion.

Galileo invented the first thermometer in the 1590s. It was an air thermometer made of a glass bulb with a long tube attached. The bulb was heated with the hands and then the tube was held vertical and the open end was partially immersed in a container with liquid in it. When the hands were removed from the bulb, the liquid in the tube rose to a certain height, and remained above the level of the liquid in the container. Although Galileo called this device a *thermoscope* rather than a thermometer, it nonetheless works sufficiently like a thermometer to have earned Galileo the title of the inventor of the thermometer.

The difference between Galileo's thermoscope and a thermometer is that a thermoscope indicates that a change in temperature has occurred, while a thermometer measures temperature change on a scale.

A Matter of Scale

Starting in the early 1700s, a number of scientists tried to devise scales to accurately measure temperature changes. Three physicists and an astronomer came up with the four temperature scales most commonly used today:

- Fahrenheit scale
- Celsius scale
- Kelvin scale
- Rankine scale

The Fahrenheit Scale: A German physicist named Gabriel Daniel Fahrenheit was the first person to use mercury as a thermometric fluid when he created his thermometer in 1714. Mercury's freezing point is substantially below that of water and its boiling point is significantly higher. Mercury also expands uniformly with changes in temperature. However, Fahrenheit did not use the freezing and boiling points of water as reference points. Instead, he used a mixture of ice, water and salt as a reference for a cold point, and the temperature of the human body as a reference for the warm point. Fahrenheit associated 0 degrees with his mixture of ice, water and salt. An ice and water mixture without the salt read 32 degrees. Fahrenheit found the temperature of the human body to be 96 degrees.

Fahrenheit's thermometer has since been recalibrated. Although 32 degrees Fahrenheit (32°F) is still recognized as the freezing point of water, the temperature of the human body is now determined to be 98.6°F and the boiling point of water – which was not acknowledged on the original scale – is now recognized as 212° on the Fahrenheit scale.

Until the 1960s, the Fahrenheit scale was the primary temperature standard for climatic, industrial and medical purposes in most English-speaking countries. In the late 1960s and 1970s, governments phased in the Celsius scale as part of the standardizing process of metrication. Today, the Fahrenheit system is the standard for non-scientific use only in the United States and a few other countries.

The Celsius Scale: A Swedish astronomer named Anders Celsius created a different scale in 1742. In this scale, 0° was the boiling point of water and 100° was the freezing point. A maker of scientific instruments later reversed these points, so the freezing point of water is 0° and the boiling point is 100°. For many years, these degrees were called "degrees centigrade." *Centi-* is a prefix meaning one-hundredth, while *grade* means degree. In 1948, the National Conference on Weights and Measures decreed that "degrees

Anders Celsius

centigrade" should be called "degrees Celsius." Today the Celsius scale is used for practically all purposes throughout the world except in the United States and a few other countries. Even in the United States, most scientists and engineers use the Celsius scale.

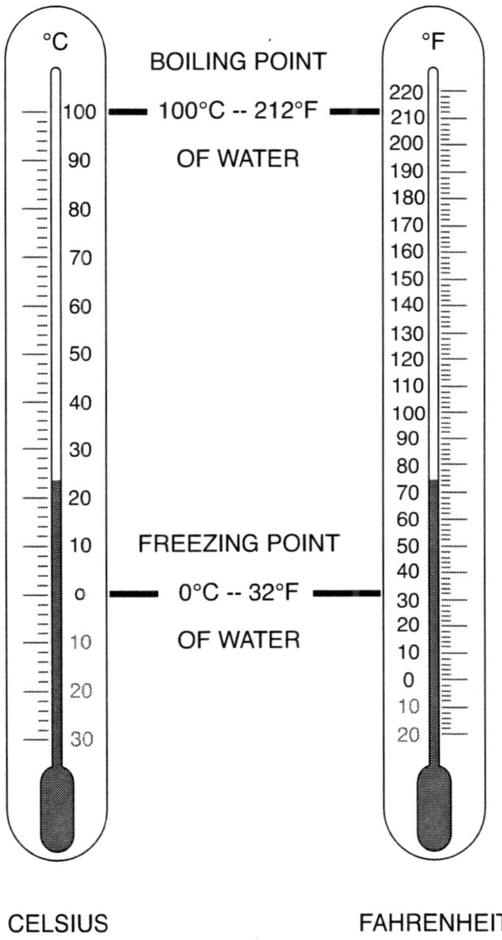

Figure 1-1. Fahrenheit and Celsius Temperature Scales ***(Courtesy of Wikipedia)***

It is possible to convert degree readings from Fahrenheit to Celsius, and vice versa. To convert from Celsius to Fahrenheit, the formula is as follows:

$$F = 9/5\ C + 32$$

To convert from Fahrenheit to Celsius, the formula is as follows:

$$C = 5/9\ (F - 32)$$

The Kelvin Scale: The Scottish physicist William Thompson introduced another temperature scale, called a thermodynamic scale because it is based on "absolute zero," in 1848. Thompson was later elevated to the rank of baron, and became Lord Kelvin. In Thompson's scale, known as the Kelvin scale, absolute zero has a value of -273.15°C. This is the temperature at which all gases, if they were to contract as much as possible, would approach the volume of zero. This is the theoretical lower limit of temperature, and at this temperature all molecular motion would cease. A temperature of absolute zero is theoretically unattainable.

To convert a temperature reading from Celsius to Kelvin, add 273.15, as shown in the following formula:

$$K = C + 273.15$$

Since water freezes at 0°C, it freezes at 273.15 Kelvin (by international agreement, the term "degree" is not used with the Kelvin scale). Since water boils at 100°C, it boils at 373.15 Kelvin. The size of one Kelvin is equal to the size of one Celsius degree.

The Rankine Scale: Another thermodynamic scale sometimes used in Great Britain is the Rankine temperature scale. William McQuorn Rankine, a Scottish civil engineer and physicist best known for molecular physics research, proposed this scale in the mid-1800s. In the Rankine scale, one unit is equal to the size of one Fahrenheit degree. A temperature of 0 degrees Rankine equals -459.67°F. To convert from Fahrenheit to Rankine, add 459.67 to the Fahrenheit value:

William McQuorn Rankine

$$R = F + 459.67$$

A few engineering fields in the United States measure thermodynamic temperature using the Rankine scale. However, throughout the scientific world, where measurements are made in SI (System International) units, thermodynamic temperature is measured in Kelvin.

Sensing the Change: Methods of Industrial Temperature Measurement

All methods of temperature measurement need to use some type of sensor. A temperature sensor has one or more properties that change in predictable ways as the temperature changes. The changing properties of the temperature sensor are interpreted as changes in temperature by a thermometer, a voltmeter or a similar device.

The five most popular types of temperature sensors – described more completely later in the chapter – are as follows:

Thermocouples: Thermocouples are the most widely used temperature sensors in industrial manufacturing environments. Thermocouples consist of two wires made of different metals that are joined at one end, called the *measurement junction*. The other ends of the wires form a *reference junction*. A current flows in the circuit when the measurement junction and the reference junction are at different temperatures. The resulting voltage is a function of the difference in temperature between the measurement and the reference junctions. The amount of voltage produced at a specific temperature depends on the types of metals used. A device that can interpret the voltage reading as a temperature value is required for measuring the temperature.

RTDs: Resistance temperature detectors, or RTDs, make use of the fact that resistance to the flow of electricity in a wire changes with temperature. Platinum is the most commonly used material for the wire.

RTDs are classified into the following two types:

- Wirewound
- Thin Film

Thermistors: Like RTDs, thermistors also change resistance with changing temperatures, but they are more sensitive than RTDs or thermocouples. Thermistors change their resistance much more significantly than RTDs with changing temperature. However, this change is highly nonlinear. Because of their extreme sensitivity and nonlinearity, thermistors are limited to measuring temperatures of a few hundred degrees Celsius. Their applications are further limited since they are less rugged than RTDs.

Infrared Thermometers: Infrared thermometers are non-contact sensors. They are used to measure temperature when contact measurement using thermocouples, RTDs or thermistors is not possible. For example, they are used to measure the temperature of moving objects, such as moving machinery or objects on a conveyor belt. They are also used where contamination is present, for hazardous conditions, or where the distance is too great for contact sensors. Infrared sensors detect the infrared energy given off by materials. The most common design includes a lens to focus the infrared energy onto a detector. The amount of infrared energy is then converted into corresponding units of temperature.

Fiber Optic Temperature Sensors: Fiber optic temperature sensors are a form of temperature measurement that uses optical fibers in making temperature measurements. Most types of fiber optic temperature sensors work by placing a temperature-sensing component on one end of the optical fiber. The other end is attached to a measuring system that collects infrared radiation and processes it into a temperature value.

Other Temperature Measurement Technologies

In addition to the above five types of temperature sensors, there are four other temperature measurement technologies that are not widely used in industrial environments. They are included here for the purpose of completeness.

Fluid Expansion Devices: The term *fluid expansion device* is a fancy name for the familiar household "mercury" thermometer and similar devices. Fluid expansion devices depend on the fact that fluids expand when heated and contract when cooled. Instead of using a separate device to interpret the liquid height and display the temperature, these thermometers contain a scale that allows the temperature to be read off directly from the height of the liquid. These thermometers do not require electric power, are not explosive, and remain stable after repeated uses. However, they must be read manually and do not generate data that is easy to record or transmit. There are other fluid expansion devices available which use gases rather than liquids.

Change of State Temperature Sensors: Change of state temperature sensors include labels, crayons, lacquers or liquid crystals, and pellets. What these devices have in common is that they change in appearance when a certain temperature is reached. In many cases, they can be used only once since this change of state is irreversible. These types of sensors are used in the food industry and are also used with steam traps.

Bimetallic Devices: Bimetallic devices are made by bonding two metal strips. They take advantage of differences in the coefficient of thermal expansion of different metals. When the strips are heated, one side expands more than the other side. As a result, bending occurs in the strips. A mechanical linkage to a pointer interprets this bending as a temperature reading. Bimetallic devices do not require a power supply. They are portable, but are less accurate than thermocouples and RTDs. Bimetallic devices are not accurate enough to record smaller temperature changes.

Integrated Circuit Temperature Sensors: Integrated circuit (IC) temperature sensors have the advantage that they are naturally linear devices; they generate an output that is proportional to the temperature. Typically, IC sensor output is stated in microamps per Kelvin. IC temperature sensors are made of semiconductor materials such as silicon and germanium; they rely on the electrical properties of these materials to produce a proportional voltage output. Voltage is converted to current using a low-temperature-coefficient thin-film resistor.

The Nitty Gritty: Technology of Industrial Temperature Sensors

As mentioned above, this section describes in some detail the technology of the following five primary industrial temperature sensors:

- Thermocouples (T/Cs)
- Resistance temperature detectors (RTDs)
- Thermistors
- Infrared (IR) sensors
- Fiber optic temperature sensors

Thermocouple Technology

A thermocouple is composed of two wires of dissimilar metals that are joined together at one end. This end, which is where the wires are exposed to the process temperature, is the measuring junction. At the other end of the wires – normally where they are attached to the measurement device – the wires form a reference, or cold, junction.

Measuring Junction

At the measuring junction, the wires are usually either soldered or welded together. There are three types of thermocouple measuring junctions:

Exposed Junction: An exposed junction has no protective assembly or covering sheath. Exposed junctions have the fastest response time, the lowest radiation error and the least conduction error. Their disadvantages are fragility and susceptibility to corrosion. Exposed junction thermocouples are also prone to pick up stray electromagnetic signals unless this is guarded against.

Grounded Junction: A grounded junction is similar to an exposed junction, except that a protective metallic sheath encloses the junction. In a grounded junction, the thermocouple wires are welded directly to the surrounding sheath material, forming a completely sealed junction. A grounded junction is more rugged and is capable of tolerating physical and mechanical abuse. It is also more resistant to corrosion and oxidation. A disadvantage of the grounded junction is slower response time. Grounded junction thermocouples are also more susceptible to conduction errors and radiation errors than are exposed junction thermocouples. Like exposed junction thermocouples, grounded junction thermocouples are also prone to picking up stray electromagnetic signals.

Ungrounded Junction: An ungrounded junction is like a grounded junction except that the junction of the thermocouple wires is not electrically connected to the metallic sheath. An electrical insulator separates the junction from the tip of the closed-end sheath. Like the grounded junction, an ungrounded junction is more rugged and tolerant of abuse. It is also shielded from electromagnetic interference. Its disadvantages are slow response time, susceptibility to conduction errors and susceptibility to radiation errors.

Measuring the Reference Junction Temperature

When the measuring junction and the reference junction are at different temperatures, a current flows from one end of the circuit to the other end. It continues to flow as long as there is a difference in temperature between the two junctions. Thomas Seebeck discovered this phenomenon in 1821, and it is called the Seebeck effect. The resulting voltage is called the Seebeck voltage. This voltage is a function of the difference in temperature between the measuring junction and the reference junction. Although it is approximately linear for small changes in temperature, it is mostly nonlinear with respect to larger changes in temperature.

Measuring the Seebeck voltage directly with a voltmeter at the reference junction does not give an accurate result because the connection between the thermocouple wires and the voltmeter leads creates a new thermoelectric circuit. In addition, since the voltage read by the voltmeter is proportional to the difference in temperature between the measuring junction and the reference junction, it is necessary to know the temperature at the reference junction to determine the temperature of the measuring junction. The output in such a measurement is usually in millivolts or microvolts.

There are several methods used to take into account the temperature of the reference junction in determining the temperature of the measuring junction, including:

Ice Bath: In the ice bath method, a junction is created within the circuit and inserted into an ice bath. This junction now becomes the reference junction, and it is at the temperature of 32°F. Since the voltage reading at the measuring junction is proportional to the difference in temperatures between the reference junction and the measuring junction, and the temperature at the reference junction is known, the temperature at the measurement junction can be determined from the voltage reading at the measurement junction.

Hardware Compensation: A second method used to take into account the temperature of the reference junction is hardware compensation. A variable voltage source is inserted into the thermoelectric circuit. This voltage source generates a compensating voltage in accordance with the ambient temperature, adding a correct voltage that cancels unwanted thermoelectric signals. Canceling these unwanted signals leaves only the voltage from the measuring junction.

Hardware compensation has the advantage that it is not necessary to actually measure the ambient temperature at the reference junction. A disadvantage of this method is that it is necessary to have a separate compensation circuit for each type of thermocouple. The circuitry used in hardware compensation also adds some error in the temperature measurement.

Software Compensation: Software compensation can be implemented when a programmable measurement system is used. In software compensation, the temperature at the reference junction is measured with a temperature sensor. Usually either a thermistor or an integrated circuit (IC) sensor is used. Once the temperature of the reference junction is known, software is used to calculate the temperature at the measuring junction. Software calculates the temperature by using the reference tables or functions built into software programs that correlate specific temperatures with voltage values for different types of thermocouples.

Thermocouple Types

Thermocouples are made up of two dissimilar metals, and they are classified according to the type of metal used to make them. Industry specifications recognize a number of types of thermocouples, which have been given alphabetical letter designations. Different thermocouple types have different temperature ranges. However, these ranges are not absolute, as other factors such as wire thickness influence the temperature range. Table 2-1 gives different types of thermocouples along with their metal composition and their temperature ranges.

Table 2-1
Thermocouple Types

T/C Type	Metal Composition	Temperature Range	Comments
J	Iron/Copper-Nickel	32°F to 1382°F	Not recommended for low temperatures; oxidizes rapidly due to presence of iron
K	Nickel-Chromium/ Nickel-Aluminum	-328°F to 2282°F	Most commonly used type; wide temperature range
E	Nickel-Chromium/ Copper-Nickel	-328°F to 1652°F	Highest voltage change per degree
T	Copper/Copper-Nickel	-328°F to 662°F	Performs well with moisture present; used for low temperature and cryogenic applications
S	Platinum-10% Rhodium/ Platinum	32°F to 2642°F	Used for very high temperatures; subject to contamination; more expensive due to noble metals
R	Platinum-13% Rhodium	32°F to 2642°F	Used for very high temperatures; subject to contamination; more expensive due to noble metals
B	Platinum-30% Rhodium/ Platinum-6% Rhodium	32°F to 3092°F	Subject to contamination; used for very high temperatures; used in the glass industry
N	Nickel-14.2% Chromium-1.4% Silicon/ Nickel-4.4% Silicon-0.1% Magnesium	-450°F to 2372°F	An alternative to type K; more stable at high temperatures

RTD Technology

The history of RTD technology goes back to Sir William Siemens. About 50 years after Seebeck's discovery concerning thermoelectricity, Siemens discovered that there is a relation between temperature change and the resistivity of metals. In making this discovery, Siemens relied on research done by Sir Humphry Davy. Siemens established the use of platinum as the element of an RTD.

The basic principle of an RTD is that as temperature increases or decreases, the resistivity of certain metals increases or decreases in a predictable and repeatable manner. The most commonly used metals are platinum, copper, and nickel. There are several reasons why these metals are used. One reason is that these metals react in a predictable way as temperature changes. Even though they do not react in a completely linear manner with temperature change, they are substantially more linear than thermocouples. Second, these metals are available in nearly pure form. Third, these metals have ductility, which means that they can all be processed into very fine wires. This is especially important when manufacturing wirewound RTDs.

There are two types of RTDs: wirewound and thin film. Wirewound RTDs consist of wire wound on a bobbin, which is enclosed in a glass or metal container. For thin film RTDs, a film is etched onto a ceramic substrate, and then sealed. RTDs are more accurate and stable than thermocouples, but cannot be used to measure temperatures higher than 660°C.

Thermistor Technology

The thermistor is another resistance-based temperature sensor. The term thermistor is derived from the phrase "thermally sensitive resistor." The resistance of thermistors also changes with changes in temperature, but the amount of change in resistance per degree is much greater in a thermistor than in an RTD. This makes a thermistor a much more sensitive device than an RTD. However, the resistance change in a thermistor is very nonlinear. As a result, thermistors are normally used only over a very small temperature span.

Thermistors are not as widely used as RTDs, and they are not very widely used in industrial applications. There are several reasons for this. One is that they have a limited span. Second, they are subject to permanent decalibration if exposed to high temperatures. Third, they are quite fragile and should not be exposed to vibration or shock. Despite their lack of popularity in industrial applications, thermistors have achieved wide use in the food transportation and service industry.

Infrared Technology

Infrared (IR) sensing is usually done by means of an infrared thermometer. An infrared thermometer measures the infrared energy emitted by materials at temperatures above absolute zero, and uses this value to determine the temperature. One basic design includes a lens that focuses infrared energy onto a detector. This energy is then converted to an electrical signal that can be displayed in temperature units. Ambient temperature variations must be compensated for to give an accurate reading. Using this arrangement, it is possible to determine the temperature of an object without making physical contact with the object being measured.

The ability to make non-contact temperature measurement of objects makes infrared technology well suited for measuring temperature in situations where probe-type sensors do not produce accurate results. Examples include objects at a distance, moving objects, objects in a vacuum, objects in an electromagnetic field or applications requiring a fast response.

While designs for IR thermometers have been around since the 19th century, the technology to create practical measuring instruments from these designs was not available until the 1930s. Since that time, many advances have been made in the use of infrared thermometers, and they have gained wide use in research and industry.

Infrared thermometers include both fixed and portable types. Portable infrared thermometers include portable infrared thermocouples and point-and-shoot infrared thermometers. Fixed infrared thermometers include fixed infrared thermocouples, online infrared thermometers and linescanners. Infrared cameras or devices based on thermal imaging technology typically are high-end infrared products, which can cost in the range of $20,000 to $100,000. These products incorporate many features that make them different from other temperature sensors.

The following sections describe the three most common types of infrared thermometers used in industry:

- Infrared thermocouples
- Portable (point-and-shoot) and fixed (online) infrared thermometers
- Infrared linescanners

Infrared Thermocouples

Infrared thermocouples, despite their name, are not actually thermocouples. Instead, they are a type of thermometer that contains an infrared detector. The output from infrared thermocouples emulates the output from particular thermocouple types. If someone wishes to replace a type K thermocouple with a non-contact form of measurement, an infrared thermocouple is available for that thermocouple type. By emulating the output from a particular thermocouple type, infrared thermocouples can replace thermocouples and provide the input that a loop controller, programmable logic controller (PLC), transmitter, or recorder is expecting. Different infrared thermocouple models are designed to match particular temperature requirements. Exergen (Watertown, MA) was among the first companies to receive a patent on infrared thermocouples, in 1992.

Many infrared thermocouples contain a sensing detector called a thermopile. A thermopile is an array of thermocouple junctions that are connected together. The thermopile contains a black material that absorbs infrared radiation. The temperature of the source is proportional to the amount of infrared radiation absorbed by the thermocouple, and the thermocouple junctions produce a corresponding voltage output. Dexter Research Center (Dexter, Michigan) is a major supplier of infrared sensing thermopile detectors. Dexter's thermopiles are hermetically sealed in an atmosphere of inert gas.

Infrared Thermometers

Infrared thermometers are available in both portable and fixed models. Portable models use a point-and-shoot method. If you point the thermometer at the material or object whose temperature you want to measure, you can read the temperature of the object on the thermometer display. Some models are available with circular laser sighting. These models show the actual area whose temperature is being measured with a red circular display. Portable models can be used to measure the temperature of many different devices. Examples of applications include measuring the temperature of electrical circuits, automobile engines, tires, concrete, steam traps, furnaces, food transportation units, heat-treating units and plastics.

Fixed infrared thermometers are also called online thermometers. Online thermometers are used to measure the temperature of materials in a fixed location, such as a process control loop. Fixed thermometers are available in a variety of body formats, operating wavelengths and output signals. Materials that are extremely hot, moving, or inaccessible are ideal candidates for online systems.

Infrared Linescanners

Infrared linescanners contain an infrared thermometer, a rotating mirror, and accompanying electronics. As the mirror scans across the product's surface, the thermometer can take a large number of individual temperature measurements. If the product is moving, two-dimensional data can be obtained. Output from the linescanner is transmitted to a personal computer, and a thermal map of the surface of the product is displayed on the computer monitor. Linescanners are used in the manufacturing of flat glass and glass windshields, and in metals manufacturing.

Fiber Optic Temperature Sensor Technology

Fiber optic temperature sensors use optical fibers in making temperature measurements. Luxtron claims to be the first company to commercialize fiber optic temperature sensor technology. Luxtron, founded in 1978, calls its pioneering temperature sensing method Fluoroptic® Thermometry. Other technologies employed by Luxtron include Radiation Thermometry and Optical Thin Film Monitoring.

It's Hard to Play Favorites: The Relative Advantages of Different Temperature Sensors

Each temperature sensor has its own merits. Thermocouples and RTDs comprise the vast majority of industrial temperature measurements, but infrared and fiber optic technology comes in handy for certain situations.

RTDs tend to respond more slowly than thermocouples. RTDs also cannot be used at high temperatures. At temperatures greater than 660°C, thermocouples or infrared thermometers must be used.

However, RTDs do have several advantages over thermocouples in industrial applications. One is that RTDs are inherently more stable than thermocouples. Typical stability for RTDs is rated at ±0.5° per year. Second, RTDs tend to be more accurate than thermocouples. The requirement that thermocouples be cold-junction compensated builds in an inherent amount of inaccuracy in thermocouples that does not have a parallel in RTDs.

Infrared thermometers can measure the temperature of objects at higher temperatures than RTDs and most thermocouples. They also can measure objects and processes at a distance, since they do not need to have contact with the object or process being measured.

In technology, linescanners fall between infrared thermometers and thermal imaging cameras. Linescanners take a large number of individual temperature measurements, and work well in measuring moving objects and processes.

Fiber optic temperature sensors function well in harsh environments, including radio frequency (RF), microwave and high voltage environments. Though they are more expensive than the other temperature sensing methods, fiber optic temperature sensors succeed in some applications where temperature cannot be measured reliably by other means.

Table 2-2 describes the advantages and disadvantages of thermocouples, RTDs, thermistors and IC sensors.

Table 2-2

Relative Advantages of Different Temperature Sensors

Quality	Thermocouples	RTDs	Thermistors	IC Sensors
Range	-400 – 4200°F	-200 – 1475°F	-100° – 500°F	-70° – 300°F
Accuracy	Less accurate than RTD	More accurate than T/C	More accurate than RTD	Highly accurate
Ruggedness	Highly rugged	Sensitive to strain & shock	Less rugged than T/Cs	Sensitive to shock
Linearity	Highly nonlinear	Somewhat nonlinear	Highly nonlinear	Highly linear
Drift	Subject to drift	Less subject to drift than T/C	Less subject to drift than T/C	More subject to drift than RTDs
Cold Junction Compensation	Required	None	None	None
Response	Fast response	Relatively slow response	Faster than RTD	Faster than RTD
Cost	Low cost except for noble metal T/Cs	Higher cost than T/C	Low cost	Low cost
Ease of Use	Application complex due to multiple types	High ease of use	High ease of use within limited parameters	Application can be complex in semiconductor applications

Temperature Units

Table 2-3 gives conversion formulas for common temperature units:

Table 2-3

Common Temperature Unit Conversions

From	To Fahrenheit	To Celsius	To Kelvin
Fahrenheit (degrees F)	°F	(F-32)*5/9	(F-32)*5/9+273.15
Celsius (degrees C)	(C*9/5)+32	°C	C+273.15
Kelvin (K)	(K-273.15)*9/5+32	K-273.15	K

Table 2-4 gives common values for different temperature measurement points:

Table 2-4
Common Temperature Measurement Values

	Fahrenheit	Celsius	Kelvin
Body Temperature	98.6°	37°	310.2
Boiling Point of Water	212°	100°	373
Freezing Point of Water	32°	0°	273
Hot Day	86°	30°	303
Cold Day	35°	2°	275
Room Temperature	68°	20°	293
Surface of the Sun	10,100°	5600°	5900

Morley's Point: Temperature

You have just finished reading the chapter on temperature measurement. I have some comments that do agree with Jesse's, but I'll reach out further into the physics of the situation.

We'll talk about temperature itself, which is the vibration of atoms – or molecules – and the energy within the group of vibrating atoms or molecules. Energy is derived from the temperature difference between groups of elements—an important difference. A lit match will not completely melt an ice cube, but a glass of warm water will.

You can think of the elements involved in a process similar to balls on a pool table. One ball does not have much energy, but many do. One fast-moving ball can share its heat by making many balls vibrate, but the group will now be composed of many balls each having a lower temperature. Lots of shaking molecules are a measure of the system entropy.

Differences in entropy are our key for doing work. Imagine, if you will, a tank of water without fish but with multiple sections and a thermal barrier between each section isolating the sections, with each section at a different temperature. Remove the thermal barriers and within a short time, the temperature becomes the same throughout. The transfer of heat can be harnessed into the definition of work in thermal considerations.

The relative energy between two segments and its transfer is your work function. Although temperature alone is an "absolute value," it is useless unless used with other temperatures. At the quantum level, temperature is much like time, a never-ceasing dance of the elements.

Several examples using the format of a fable might help. In my younger days I designed some systems for buildings and building controls. Temperature and comfort were key elements in these designs. This was in the early 1980s. The team considered the various offerings for temperature sensing and chose thermistors. Why? They were repeatable, low-cost and had their highest sensitivity at around 100°F – ideal for building temperature control. My science friends insisted upon thermocouples. If you look at Jesse's comparison, you see that you have to be a scientist to make a thermocouple work. Temperature control worked like a charm using thermistors. To my surprise, we could also use them to measure flow. The temperature differential across some copper tubing is an indicator of flow. Not "accurate," but I'm just an engineer.

I once set up an outdoor hot tub for my beautiful wife, located right outside the master bedroom. To my surprise I found that she wanted the temperature higher in the summer and lower in the winter. Her preference was not dependent upon the outside temperature, but upon the seasons themselves. The user defines the specifications.

The following example uses a surprising effect of temperature. You can do this experiment at home. I had an iron frying pan with a metal handle and had cooked some bacon. I put it under cold water and found that an instantaneous flash of heat came up the handle. That's not supposed to happen — thermodynamics forbids it. I realized that the time interval of measurement is very important.

Another example of the importance of the time interval of measurement is the proving or disproving of the principles of "cold fusion." At a conference on this subject, the participants discovered that the temperature was not being measured with the right time considerations. Most temperature measurements are made over time and are not instantaneous. When measuring instantaneous temperature, you may get instantaneous (but misleading) peak temperatures.

Jesse has done a good job of erecting the "scaffold of usability" of temperature measurements, definitions and best practices. Understanding temperature is key to making some critical decisions. Come on in, the water's fine.

"Pressure makes diamonds."
—General George S. Patton

Chapter Three

Measurement Under Pressure

Pressure is a state or condition we face every day. We speak of being "under pressure" and talk about putting pressure on another person or situation. Fundamentally, the idea of pressure is the idea of the continuous exertion of force. From a measurement perspective, pressure is force per unit area. The presence of atmospheric pressure also plays an important role in pressure measurement by pressure transmitters. In process control, pressure measurement occurs in a variety of contexts, especially including the measurement of the pressure exerted by liquid, steam and gas.

General George S. Patton

What Is Pressure?

The *American Heritage Dictionary* has the following definitions of *pressure.*

As a noun:

1. a. The act of pressing.
   b. The condition of being pressed.
2. The application of continuous force by one body on another that it is touching; compression.

3. *Abbr.* **P** *Physics.* Force applied uniformly over a surface, measured as force per unit of area.
4. *Meteorology.* Atmospheric pressure.
5. A compelling or constraining influence, such as a moral force, on the mind or will: *pressure to conform; peer-group pressure.*
6. Urgent claim or demand: *under the pressure of business; doesn't work well under pressure.*
7. An oppressive condition of physical, mental, social, or economic distress.
8. A physical sensation produced by compression of a part of the body.
9. *Archaic.* A mark made by application of force or weight; an impression.

As a verb:

1. To force, as by overpowering influence or persuasion.
2. To pressurize.
3. To pressure-cook.

This chapter focuses mainly on the following elements of pressure measurement:

- History of pressure measurement
- Pressure sensing technologies
- Pressure transmitters
- Differential pressure flowmeters
- Primary elements
- The future of pressure measurement

Conversion values for common units of pressure are given at the end of the chapter.

Pressure Transmitters Feel the Pressure

Pressure transmitters assume the existence of pressure, and transmit this pressure in quantitative form to an instrument, device or controller that is capable of acting on this value. Pressure transmitters include the following elements:

- Pressure sensor
- Transducer, amplifier or conditioning element
- Output

Pressure transducers, which are also sold separately, are generally lower in cost and smaller than pressure transmitters, and are typically not used in the process industries. They typically have loose wires at one end, and do not perform at the same level as pressure transmitters.

Four Types of Pressure Transmitters

Pressure transmitters measure the pressure exerted by liquid, steam or gas. There are four main kinds of pressure transmitters:

- Multivariable (MV) pressure transmitters measure two or more process variables in a single device. These are usually pressure and temperature.
- Differential pressure (DP) transmitters measure a difference in pressure upstream and downstream of a constriction in the pipe, called a *primary element.*
- Gage pressure transmitters measure an amount of pressure that includes atmospheric pressure.
- Absolute pressure transmitters measure an amount of pressure that does not include atmospheric pressure.

Because of the importance of pressure in flow measurement, this chapter mainly focuses on differential pressure transmitters and on primary elements. Primary elements are used with pressure transmitters to measure flow.

From Roman Nozzles to Stolz's Universal Orifice Equation: How Pressure Measurement Evolved

The history of differential pressure (DP) flow measurement goes back to at least the 17th century, though the measurement of flow using nozzles goes back to Roman times, in the 2nd century A.D. At the beginning of the 17th century, Torricelli and Castelli arrived at the concepts that underlie differential pressure measurement today: that flowrate equals velocity times pipe area, and that the flow through an orifice varies with the square root of the differential pressure. In 1738, Bernoulli developed his famous equation for flowrate calculation: $V^2/2g + z + P/\rho g = \text{Constant}$.

The development of primary elements for use in measuring DP flow also began about this time. The Pitot tube is named for Henri Pitot, who invented it in 1732. Henry Philibert Gaspard Darcy, another Frenchman, authored a paper that made improvements on Pitot's invention. This paper was published in 1858, shortly after his death. (Brown, G. "Henry Darcy and the Pitot Tube." *International Engineering History and Heritage* (2001): pp. 360–366) The first patent for the use of a Pitot tube to measure velocity in pipes was given to Henry Fladd of St. Louis, Missouri in 1889.

The Venturi tube was invented by an Italian physicist named Giovanni Battista Venturi in 1797. In 1887, Clemens Herschel used Venturi's work to develop the first commercial flowmeter based on it. His version of the Venturi flowmeter became known as the Herschel Standard Venturi. Herschel published his paper called "The Venturi Water Meter" in 1898. In 1970, a company called BIF introduced the Universal Venturi Tube™.

Max Gehre received one of the first patents on orifice flowmeters in 1896. The first commercial orifice plate flowmeter appeared in 1909, and was used to measure steam flow. Shortly thereafter, the oil and gas industries began using orifice plate flowmeters due to their ease of standardization and low maintenance. In 1912, Thomas Weymouth of the United Natural Gas Company did experimental work on the use of orifice flowmeters to measure natural gas flow. Weymouth used pressure taps located one inch upstream and one inch downstream of a square-edged orifice. The Foxboro Company licensed Weymouth's work and used it as a basis for building orifice meters shortly after this time.

The increased use of orifice meters captured the attention of several engineering organizations. These included the American Gas Association (AGA), the American Petroleum Institute (API) and the American Society of Mechanical Engineers (ASME). The National Bureau of Standards (NBS) also became involved in this research. In 1930, a joint AGA/ASME/NBS test program was able to generate a coefficient-prediction equation based on extensive tests. In 1935, tests performed at Ohio

State University in conjunction with the National Bureau of Standards served as the basis for flow equations that have been used by the AGA and the ASME since that time.

Work in the United States was combined with European work in the late 1950s, and resulted in the issuance of ISO Standards R541 for orifices and nozzles and R781 for Venturis. Standard R541 was published in 1967 and R781 was published in 1968. At about the same time, an ASME Fluid Meters Research Committee began a study to re-evaluate the Ohio State University data and to add new data on coefficients. The results were issued in an ASME Fluid Meters Report in 1971.

J. Stolz proposed a universal orifice equation in 1975. His idea was to combine the Ohio State University data into a single dimensionless equation that could be used for corner, flange and D- and D/2 taps. He presented his equation in a paper in 1978. This equation appears in ISO Standard 5167, published in 1980, which combines the previously published R541 and R781 standards into a standard for DP flow. The ASME Fluid Meters Research Committee adopted the ISO 5167 standard in 1981. In 1995, this standard was developed into the ASME MFC-3M standard for all orifice, Venturi and nozzle flowmeters.

Piezoresistive Sensors Lead the Pressure Sensing Technologies

The following are pressure sensing technologies used in pressure transmitters:

- Piezoresistive sensor
- Strain gage
- Capacitive sensor
- Other

Of these four categories, piezoresistive is the leading sensing technology in terms of use, followed by strain gage.

Piezoresistive Sensors

Piezoresistivity is a property of a material in which its electrical resistance changes when it is subjected to an external pressure. Piezoresistive pressure sensors typically contain several thin wafers of silicon that are embedded between protective surfaces. The surfaces are connected to a Wheatstone bridge (a circuit widely used to read the sensor output precisely), which is able to detect differences in resistance. A small amount of current passes through the pressure sensor at a particular pressure. Its resistance changes as the applied pressure changes and so does the current. This difference is detected by the Wheatstone bridge, and a change in pressure is reported. A diagram of a Wheatstone Bridge is shown in Figure 3-1.

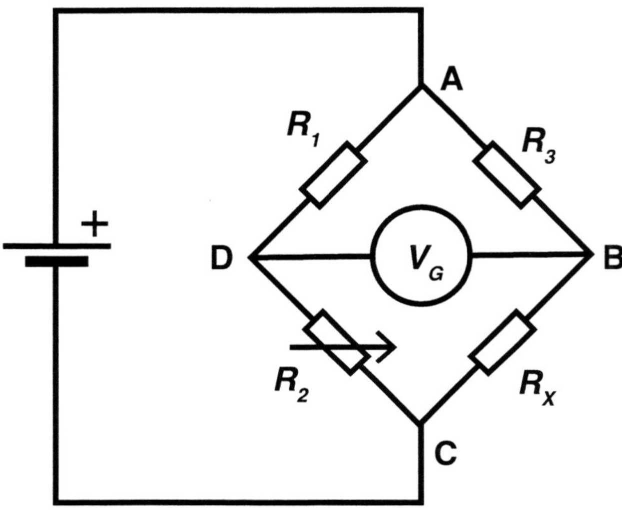

Figure 3-1. Wheatstone Bridge

Strain Gages

Strain gages are sensors whose resistance changes when they are strained. Strain gage pressure sensors measure the displacement of an elastic diaphragm as a result of a pressure differential across the diaphragm. In a strain gage sensor, the strain gage is bonded to a diaphragm and wired into a Wheatstone bridge configuration. Strain generates a change in electrical resistance that is proportional to the change in pressure.

Capacitive Sensors

Capacitive pressure sensors have a thin diaphragm as one of the two plates of a capacitor. The diaphragm is exposed to a reference pressure on one side and the process pressure on the other. Changes in pressure experienced by the diaphragm change the distance between the plates, causing a change in the capacitance of the sensor. While stainless steel is the most commonly used diaphragm material for capacitance sensors, Inconel and Hastelloy also give good results. Tantalum is used for high temperature applications.

Other Types

Other types of pressure sensors include potentiometric and resonant wire. Potentiometric sensors have a potentiometer with a wiper arm that is linked to a Bourdon or bellows element. As the wiper arm moves across the potentiometer, the pressure sensor deflection is changed into a resistance measurement with a Wheatstone bridge circuit. A resonant wire sensor contains a wire that is held by the sensing diaphragm at one end and a static member at the other end. The wire oscillates at resonant frequency, due to an oscillator circuit. As the process pressure changes, the tension in the wire changes proportionally, resulting in a change in the resonant frequency of the wire.

What Is a Differential Pressure Flowmeter?

When most flowmeters are sold, the transmitter and sensor are sold together. This is true for ultrasonic, vortex, Coriolis, turbine, and other types of flowmeters. All of these flowmeters operate based on a correlation between flowrate, or mass flow, and some physical phenomenon. For ultrasonic flowmeters, it's the difference in transit time of sound waves sent across the pipe. For turbine flowmeters, it's the speed of the rotor. Differential pressure flowmeters also correlate flow with a physical phenomenon: the difference in pressure upstream and downstream from a constriction in the flowstream.

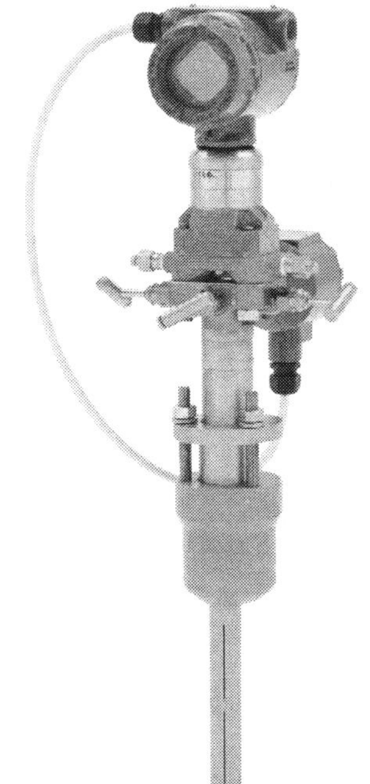

Figure 3-2. An Integrated DP Flowmeter ***(Courtesy of Emerson Process Management)***

Where differential pressure flowmeters differ from other flowmeter types is that the transmitter is often sold separately from the primary element that creates the constriction in the flowstream. In many cases, the primary element is bought from a different supplier than the pressure transmitter supplier. Because a differential pressure flow transmitter cannot make a flow measurement without a primary element, customers who buy a DP flow transmitter without the primary element are not actually buying a flowmeter. They don't have a DP flowmeter until they connect the primary element to the DP flow transmitter. A DP flowmeter, then, is considered to be a DP flow transmitter that is connected to a primary element for the purpose of making a flow measurement.

Energy Conservation – The Theory of Differential Pressure Measurement

The theory behind DP flowmeters is that energy is conserved when flow passes across or through a constriction in the pipe. A more exact statement of this theory is known as Bernoulli's principle, which states that the sum of the fluid's static energy, kinetic energy and potential energy is conserved across a constriction in the pipe. One form of Bernoulli's principle for incompressible fluids is as follows:

Daniel Bernoulli, author of the Bernoulli principle

$$V^2/2g + z + P/\rho g = \text{Constant}$$

Here V is the velocity of the fluid, g is the acceleration due to gravity, P is pressure, ρ is the density of the fluid and z is the elevation head of the fluid.

The equation of continuity formulates a relation between fluid flowrate and velocity for fluids that are incompressible. It can be formulated as follows:

$$Q = A_1 \times v_1 = A_2 \times v_2$$

Here Q = volumetric flowrate

A_1 is the cross-sectional area of the pipe upstream from the constriction in the pipe.

v_1 = flow velocity upstream of the constriction in the pipe.

A_2 is the cross-sectional area of the pipe downstream from the constriction in the pipe.

V_2 = flow velocity downstream of the constriction in the pipe.

Combining Bernoulli's principle with the equation of continuity yields the result that the differential pressure generated by an orifice plate is proportional to the square of the flow through the orifice plate. This is the "square root" relationship that is fundamental to all orifice plate and other DP flow measurements.

The square root relationship that is fundamental for DP flow measurement is shown in the following flow equation:

$$q_m = \frac{C\,Y\,\pi\,d^2\sqrt{2g_c\ \rho_f\ \Delta P}}{4}$$

Here q_m = mass flowrate

C = discharge coefficient of flow at the Reynolds Number

Y = expansion factor (usually 1.0 for liquids)

π = pi

d = constriction diameter at flowing conditions

g_c = gravitational attraction

ρ_f = fluid density

ΔP = differential pressure

Some companies have brought out integrated products that include the primary element mounted together with a DP flow transmitter. An example includes Emerson's Rosemount 3051SFA Annubar Flowmeter, pictured in Figure 3-2. It is tempting to consider these as the only true DP flowmeters; however, a better description is that these are DP flowmeters with an integrated primary element. If a customer assembles a DP flowmeter by connecting up a DP flow transmitter to an orifice plate or a Venturi tube from another source, the result is just as much a flowmeter as an integrated product.

As has been mentioned, a DP flowmeter has two separate components that must be brought together to create a DP flowmeter: a DP transmitter and a primary element. To better understand these two components, this book focuses on them separately. With the discussion of DP transmitters complete, it is time to focus on primary elements.

Primary Elements – Not Glamorous, but Essential

Primary elements are an essential component of DP flow measurement. Primary elements place a constriction in the flow line that creates a pressure drop in the line. A DP transmitter uses the difference between upstream pressure and downstream pressure in the line as a basis for computing flow. While absolute and gage pressure transmitters only measure pressure, multivariable and DP transmitters measure differential pressure across primary elements in order to measure flow. The following types of primary elements are included here:

- Orifice measuring points
- Venturi tubes
- Averaging Pitot tubes
- Flow nozzles
- Wedge elements
- Other primary elements

Orifice Measuring Points

An orifice measuring point includes an orifice plate, an orifice assembly, flange, or holding element, and in most cases, a valve manifold. Details on these follow:

Orifice plates are the most common type of primary element. An orifice plate is a flat, usually round piece of metal, often steel, with an opening or "orifice" in it (Figure 3-3). The orifice plate needs to be positioned at a correct location and orientation in the flowstream for it to function as a primary element for the purpose of making a differential pressure flow measurement. For it to be so positioned, it must be held in place. This is typically done by an orifice assembly, an orifice flange (Figure 3-4) or a holding element.

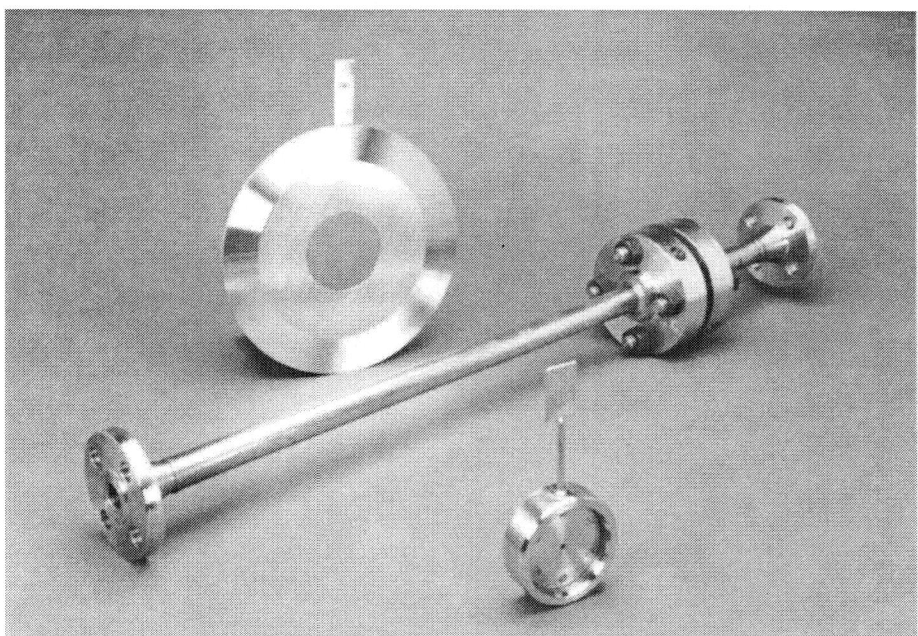

Figure 3-3. Orifice Plates *(Courtesy of ABB)*

In addition to an orifice plate and assembly or flange, most orifice plate installations require the presence of a valve manifold, which serves to isolate the pressure transmitter from the process. DP flow transmitters use either a three-valve or a five-valve manifold.

Orifice plates are classified according to the shape and position of the hole or opening they contain. The following are the main types of orifice plates:

- Concentric
- Conical
- Eccentric
- Integral
- Quadrant
- Segmental

Figure 3-4. Orifice Flange Assemblies *(Courtesy of ABB)*

Pitot Tubes

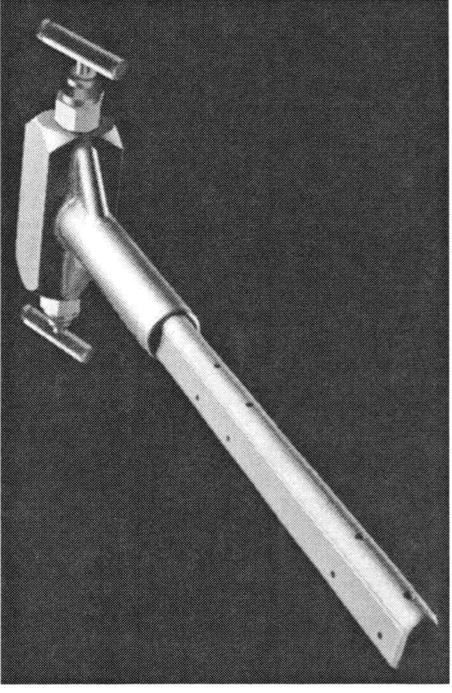

Figure 3-5. Verabar Multiport Averaging Pitot Tube *(Courtesy of Veris, Inc.)*

Pitot tubes use a difference in pressure as a basis for computing flowrate. They are of two types:

- Single port
- Multiport averaging Pitot tubes

A **single port Pitot tube** includes an L-shaped tube measuring impact pressure. This tube is inserted into the flowstream, with the opening facing directly into the flow. Another tube, measuring static pressure, has an opening parallel to the direction of flow. Flowrate is proportional to the difference between impact pressure and static pressure.

A **multiport averaging Pitot tube** (Figure 3-5) has multiple ports to measure impact pressure and static pressure at different points. The DP transmitter computes flowrate by taking the average of the differences in pressure readings at different points. This gives a more accurate reading than measuring flow at a single point.

Some companies such as Emerson Process Management and Veris have introduced proprietary versions of the averaging Pitot tube. Emerson's Rosemount proprietary version is called the Annubar, and it was formerly sold by Dieterich Standard, now part of Emerson Process Management. Veris' averaging Pitot tube is called the Verabar.

Venturi Tubes

A **Venturi tube** (Figure 3-6) is a flow tube that has a tapered inlet and a diverging exit. The DP transmitter measures pressure drop and uses this value to calculate flowrate.

Figure 3-6. Venturi Tubes *(Courtesy of ABB)*

Flow Nozzles

A flow nozzle (Figure 3-7) is a flow tube with a smooth entry and a sharp exit. The DP transmitter computes flowrate based on the difference between upstream pressure and downstream pressure. Flow nozzles are mainly used for high-velocity, erosive, non-viscous flows. Flow nozzles are sometimes used as an alternative to orifice plates when erosion or cavitation would damage an orifice plate. They offer excellent long-term accuracy.

Figure 3-7. A flow Nozzle *(Courtesy of ABB)*

Wedge Elements

A wedge element is a flow tube that has a V-shaped flow restriction protruding into the flowstream from at least one side of the pipe (Figure 3-8). Wedge elements are designed to measure fluids with a high solids content. For example, slurries and many wastewater liquids have solids in them. Because the liquid does not go through a narrow opening, wedge elements can easily handle these flows. They are also well-suited for air and viscous flows.

Figure 3-8. COIN Wedge Meter (Courtesy of BadgerMeter)

Other Primary Elements

Other primary elements include low loss flow tubes, Dall tubes, and the V-Cone primary element. Low loss flow tubes are designed to produce a minimum amount of permanent pressure loss. The Dall tube was invented by an ABB hydraulics engineer named Horace E. Dall. It is an adaptation of the Venturi tube. The V-Cone is a proprietary device that is designed for flow measurement with minimal upstream piping. It is manufactured and sold by McCrometer in Hemet, California.

One other category that deserves mention here is laminar flow elements. They are often used for air and gas flow measurement. Laminar flow elements are used with mass flow controllers to create a pressure drop and a flow measurement. They are also used to measure air flow to internal combustion engines.

Advantages and Disadvantages of Various Primary Elements

Table 3-1 shows the advantages and disadvantages of the various primary elements. It also describes the applications and the typical line sizes.

The Future of Pressure Measurement

Pressure sensors, transducers, and transmitters are widely used in industrial and manufacturing operations worldwide. The value of the pressure transmitter market worldwide exceeded $3 billion in 2013. This market will continue to grow as new applications emerge and suppliers bring out new products.

During the past five years, pressure transmitter suppliers have made substantial strides in improving the accuracy and long-term reliability of pressure transmitters. They have also made improvements to multivariable pressure transmitters and to the self-diagnostic capabilities of pressure transmitters. Being able to anticipate or diagnose failures electronically reduces the need for manually checking on these transmitters. Expect to see a steady stream of innovations from pressure transmitter suppliers in the future.

Table 3-1
Advantages and Disadvantages of Different Primary Elements Used with DP Flowmeters

Flowmeter Type	Advantages	Disadvantages	Liquid, Steam, or Gas	Line Size	Comment
DP-Orifice Plate	Low initial cost Ease of installation Well understood	Limited range Permanent pressure drop Uses square root method to calculate flowrate	Liquid, Steam, Gas	½ inch and up	Most traditional meter
DP-Venturi Tube	Suitable for clean and dirty liquids	Can be unwieldy and difficult to install due to size	Liquid, Gas	2–30 inches	Greater size for gas
DP-Pitot Tube	Low cost Virtually no pressure drop	Low accuracy Limited sampling	Liquid, Gas	> 1 inch	Measures only at a single point
DP-Averaging Pitot Tube	More accurate than single Pitot tube Virtually no pressure drop	Limited range Not suitable for dirty fluids	Liquid, Gas	> 1 inch	Available as Annubar
DP Flow Nozzle	Good for high velocity fluids Handles dirty fluids better than orifice plate	High initial cost Difficult to remove for inspection and cleaning	Steam	2–30 inches	Used for steam applications

Units of Pressure

Table 3-2 provides conversion values for some of the common units of pressure.

Table 3-2
Conversion Values for Common Units of Pressure

Pressure Unit	Abbreviation	Conversion
Millimeters of Mercury	1 mmHg	133.322368 Pascals
Pascal	1 Pa	0.007501 mmHg
Millimeters of Mercury	760 mmHg	101.325 kPa
Pascal	1 Pa	1 N/m^2
Newton per Square Meter	1 N/m^2	1 Pa
Atmosphere	1 atm	760 mmHg
Atmosphere	1 atm	14.696 psi
Pounds per Square Inch	1 psi	6894.75 Pa
Pounds per Square Inch	1 psi	0.068046 atm
Pounds per Square Inch	150 psi	10.206884 atm
Pounds per Square Inch	1 psi	0.0689475 bar

PSIG and PSIA

The unit psi is widely used in the United States to measure pressure in instruments. Variations on this unit are psig and psia. The unit "psig" means "pounds per square inch gage," which is derived from pressure read from a gage. A gage in a pipe reads the difference between the pressure in the pipe and atmospheric pressure. "psia" means "pounds per square inch absolute" and means that the pressure includes atmospheric pressure. Therefore, psia is higher than psig. Their relationship is as follows:

psig + 1 atm = psia
psia – 1 atm = psig

As an example, consider a car tire pumped up to 30 psi above atmospheric pressure. This tire will have 30 + 14.7 = 44.7 psia or 30 psig.

Morley's Point:
Pressure

Jesse really covers ground where pressure is related to flow measurement. I'd like to expand the concept of pressure into different areas. To do this, first I had to get my brain back in gear. I looked at the MIT course in fluid mechanics and went through it for an afternoon. It made me feel old and young at the same time and brought memories flooding back — I hadn't forgotten it all.

Pressure is defined as force per unit surface area. This means that if you increase surface area and maintain a constant force you will reduce pressure (and conversely). This is the principle of the hand-powered hydraulic jack. A given force on a small area that is connected with oil to a large area results in a proportionally larger force being exerted by the large area.

But there are other forms of pressure such as marital spouse arguments. Let's forget that along with flow for the time being. And the other meanings of "pressure" in language – job pressures, etc. – are not a concern here. (Besides, we get enough of that kind of pressure.) The printing process was one of the first uses of pressure. You make a model of the page using protruding typefaces, which were originally made from wood and then lead, then apply ink to the typefaces and press paper against the ink-covered surfaces. That's called pressing.

Knowing how pressure works allows us to manufacture and mold materials such as clay and metals through the smart manipulation of forces and pressure. Just for the heck of it, I looked up the design of the chicken egg. Evolution has designed the egg to accept pressure without breaking. I believe the swan's egg is good for 26 pounds of pressure before cracking, and the ostrich's is good for more like 120 pounds — pretty good for calcium carbonate.

My first real contact with flow dynamics was in the mid-1950s. I was working on a satellite program and they needed some form of memory for a satellite. (We must remember that these times were long ago and far away.) We had to put up an imaging satellite immediately after flights over Russia were banned. Russia had put up Sputnik and we were concerned. In maybe a year, according to estimates, they would be able to toss bombs out of a satellite. We had no satellites capable of that. What should we do?

Our team came up with a solution that we called the Bernoulli disk. Why a disk? It would only rotate when you needed to read or write, saving power. It had to survive immense vibration during takeoff and have consistent memory performance independent of vibration or temperature. What could we do? We know that an increased flow reduces pressure so we cut a circular magnetic disk from tape and put special cubic magnetic material on the disk surface. Read/write heads were fixed in a plate under the disk and we filled the container with nitrogen.

A constant separation between the disk and heads was required even under severe G forces. Acting as a centrifugal pump, the flow was stable and the distance was a constant. The design was vibration and temperature tolerant and had very low power demand because it only operated when the read/write cycle was used. Put mobile heads on this and call it a floppy disk. A playing card skimming across the table was the inspiration for this. Those were the days back in 1958. The disk never made it to a satellite, however, as it was backup technology. The technology chosen was a standard film camera; the satellite ejected the exposed film with a parachute, which was then caught midair. Anyway, it worked – most of the time.

In the future of pressure and fluid dynamics, there is lots to be done: the new fans with no blades, windmills that "work," centrifugal filters that have no filter and supersonic aircraft. All applications need a sophisticated analysis of fluid dynamics to make them work. While doing the research for this chapter, I stumbled across an interesting fact about how we can now position, with lasers, material in space. Using pressure is not the only way to measure flow. So much to do, and so little time!

"Everything flows and nothing abides, everything gives way and nothing stays fixed." – Heraclitus

Chapter Four

Flow Measurement – How Do You Measure Continuously Moving Flow?

Flow is a commonly used word, but most people would be hard-pressed to define it precisely. One way to look at flow is to say that it is the natural state of a fluid. But what does this actually tell us about flow? Scientists divide matter into three states: solids, liquids and gases. Liquids and gases are both fluids, so we could just as easily divide matter into two states: solids and fluids.

The key idea behind flow is the idea of continuous movement. But a baseball hit for a home run in the center field seats doesn't flow into the seats; it flies. And a jogger who runs five miles is in continuous motion, but he doesn't flow; he merely runs or jogs. Perhaps this is because a baseball and a human body are objects, from the point of view of physics. But even objects can flow under certain conditions – we say, for example, traffic flows. The difference in this case is that a group of objects are moving continuously in some kind of pattern.

Since the idea of flow is so closely connected to fluids, it's better to look back at fluids rather than to derive the idea of flow from the continuous movement of solids. **Flow is the natural state of fluids when they are moving continuously in some direction**. The ocean lapping at the shore isn't flowing because this is more of a back and forth movement than movement in a single direction. But a river or stream flows to the sea and water flows through pipes on the way to being consumed at a house. The movement of open channel flow is due to gravity, while flow in pipes is due to pressure. This is a broad definition, but it accounts intuitively for many of the most common examples of flow.

The idea of continuity has presented challenges to mathematicians and philosophers for many years. Flow and continuity are closely related; they are not quite synonymous, but it is not possible to subtract the idea of continuity from flow. Unfortunately, many mathematicians have chosen to analyze continuous phenomena in terms of a large number of discrete points or objects. For example, the number line, which is continuous, is viewed as a large number of discrete points. Since there never seem to be enough points to make up a continuous line, the idea of infinity is used as a kind of cosmic glue: the number line is made up of infinitely many discrete points. These subjects will be discussed more in Chapters Five and Six.

Flow Measurement Is Vital to Water & Wastewater, Oil & Gas, and Other Industries

There are many occasions when fluid is moving from one location to another that it has to be measured. **Flow measurement is measuring the quantity of fluid that flows through a pipe or open channel in a period of time. The quantity can be volumetric or gravimetric (mass). The fluid can be air, liquid, or gas. Even solid particles can be measured as a fluid, if they move continuously in a direction as a group.**

Most of this chapter is devoted to discussing flow measurement. Flow measurement is an important topic in industrial environments because entire industries are built around the use and movement of certain fluids. For example, the water & wastewater industry is devoted to the movement and correct use of clean water for drinking and human consumption, and to the correct movement and disposal of used water in the form of wastewater. The chemical and food & beverage industries rely on accurate measurement of both liquids and gases in their manufacturing processes. The oil & gas production and transportation industry is devoted to finding crude oil, separating it into its components, and moving it to where it can best be used. This may be a refinery or storage area. Once oil is refined into petroleum liquids, these liquids are distributed at downstream locations. And many of these measurements involve custody transfer, which requires a high degree of accuracy.

Measuring the quantity of fluid that flows makes it possible for ownership of the fluid to change from one person or company to another. This is called *custody transfer*. It also makes it possible in a manufacturing environment, such as a chemical plant, to follow a recipe for a product by controlling how much of each liquid or gas goes into making the product. Clearly, flow measurement is vitally important in a variety of contexts.

New and Traditional Technology Meters Battle to Measure the Flow

Flow measurement is done by flowmeters. **A flowmeter is a device designed to measure the quantity of fluid flow in a closed pipe or an open channel**. The rest of this chapter discusses 12 different types of flowmeters, their origin, principle of operation, paradigm case application, and applications.

One of the most interesting developments in the flowmeter market today is the battle between the newer flow technologies and traditional flowmeters. New-technology flowmeters include Coriolis, magnetic, ultrasonic, vortex and thermal flowmeters. Traditional flow technologies include differential pressure (DP), turbine and positive displacement meters. While there is a general trend toward the new-technology meters and away from the traditional meters, this change varies greatly by industry and application.

When users select flowmeters today, they are faced with a variety of choices. Not only are many technologies available, but there are also many different suppliers for each technology. When ordering replacement meters, users often replace like with like. This is one reason that DP flowmeters still have the largest installed base of any type of flowmeter. In other cases, users need to select meters for new plants, or for new applications within existing plants. Users also sometimes replace one type of flowmeter with another type. How should this selection be made?

This chapter addresses the issue of flowmeter selection by examining the operating principles, advantages and limitations of new-technology flowmeters. It then presents a step-by-step method of flowmeter selection that takes these factors into account, along with application, performance, cost and supplier conditions. This method is called a "paradigm case" method for selecting flowmeters.

Because the first step in the paradigm case method of flowmeter selection involves selecting flowmeters whose paradigm case application matches a particular application, we have identified the paradigm case application for each type of flowmeter. A paradigm case application is one in which the conditions are optimal for the operation of that type of flowmeter. The paradigm case application for each flowmeter type is determined by the physical principle that underlies the flowmeter technology.

New-technology flowmeters have not been around as long as traditional flowmeters, but their performance level is typically higher. They have the following characteristics:

1. They were introduced after 1950.
2. They incorporate technological advances that avoid some of the problems with earlier flowmeters.
3. They are more the focus of product development by suppliers than traditional flowmeters.
4. Their performance level, including criteria such as accuracy and reliability, is at a higher level than that of flowmeters introduced earlier.
5. Manufacturers of these meters have been quick to adopt communication protocols such as HART, FOUNDATION Fieldbus and Profibus.

Traditional technology flowmeters, by contrast, have been around for many years. They have the following characteristics:

1. They were introduced before 1950.
2. They have higher maintenance requirements than new-technology flowmeters.
3. They are less the focus of product development by suppliers than new-technology flowmeters.
4. Their performance level, including criteria such as accuracy and reliability, is at a lower level than that of new-technology flowmeters.

5. Manufacturers of these meters have been slow to adopt communication protocols such as HART, Foundation Fieldbus, and Profibus.

New-Technology Flowmeters Emerge with the Baby Boom

New-technology flowmeters are so-called because they represent technologies that have been introduced more recently than DP flowmeters and other traditional flowmeters. Most of the new-technology flowmeters came into industrial use in the 1960s and 1970s, while the history of DP flowmeters goes back to the early 1900s. Each new-technology flowmeter is based on a different physical principle, and represents a unique approach to flow measurement.

The new-technology flowmeters (Coriolis, magnetic, ultrasonic, vortex, and thermal meters) have all been introduced since 1950. Magnetic flowmeters were first introduced in the Netherlands by a company called TOBI in 1952. Tokimec first introduced ultrasonic meters in Japan in 1963. Eastech brought out vortex flowmeters in 1969, while Coriolis meters came onto the market in 1979. Thermal flowmeters were developed in the mid-1970s.

Coriolis Flowmeters Twist the Flow

Coriolis flowmeters are named after the French mathematician Gustave Coriolis. In 1835, Coriolis showed that an inertial force must be taken into account when describing the motion of bodies in a rotating frame of reference. The rotation of the earth is commonly used to illustrate this Coriolis force. Because the earth is rotating, an object thrown from the North Pole to the equator will appear to deviate from its intended path.

Coriolis flowmeters are composed of one or more vibrating tubes, usually bent. The fluid to be measured passes through the vibrating tubes. The fluid accelerates as it passes toward the point of maximum vibration and decelerates as it leaves this point. The result is a twisting motion in the tubes. The degree of twisting motion is directly proportional to the mass flowrate. Position detectors sense the positions of the tubes. While most Coriolis flowmeter tubes are bent, and a variety of designs are available, some manufacturers have also introduced straight tube Coriolis flowmeters.

It is often said that Coriolis flowmeters measure mass flow directly, unlike other flowmeters that calculate mass flow by using an inferred density value. Volumetric flow (Q) is calculated by multiplying the cross-sectional area (A) of a pipe times the average fluid velocity. Mass flow is determined by multiplying volumetric flow (Q) times the density (ρ) of the fluid. Some multivariable flowmeters measure the pressure and temperature of the process fluid, and then use these values to infer fluid density. Mass flow can then be calculated.

One way that Coriolis suppliers differentiate themselves is by tube design. Most suppliers have proprietary designs for the bent tubes that make up the Coriolis meter body. Some companies offer single bent tubes, while others have dual bent tubes. In 1987, Endress+Hauser introduced the first dual tube straight-tube Coriolis flowmeter. In 1994, KROHNE became the first company to introduce a single tube straight-tube Coriolis meter. Straight-tube meters are easier to clean than bent-tube meters, and also have less pressure drop.

Industry Approvals Promote Coriolis Usage

One important function of the American Gas Association (AGA) and the American Petroleum Institute (API) is to formulate criteria or standards for buyers and sellers to follow when transferring ownership of natural gas and petroleum liquids. In the past, these organizations have published reports on the use of differential pressure and turbine flowmeters for use in custody transfer of natural gas.

Industry approvals have played a role in the growth of the Coriolis flowmeter market. The AGA published a report named AGA-11 in 2003, outlining criteria for using Coriolis flowmeters for custody transfer of natural gas. The API published a draft standard called "Measurement of Single-Phase, Intermediate, and Finished Hydrocarbon Fluids by Coriolis Meters" in 2000. The API approved a second draft standard called "Measurement of Crude Oil by Coriolis Meters" in 2001.

Paradigm Case Application

The paradigm case application for Coriolis meters is with **clean liquids and gases flowing sufficiently fast to operate the meter and flowing through pipes of size**

four inches or less. It is also important that a mass flow rather than a volumetric flow measurement is desired. The primary limitation on Coriolis meters is size, since they become quite unwieldy and expensive in line sizes over four inches. Some low-pressure gases do not have sufficient density to operate the flowmeter. One advantage of Coriolis meters is that the same flowmeter can be used to measure different types of fluids, including some fluids with varying densities that cannot be measured by other flowmeters. Coriolis meters can measure the mass flow of slurries and dirty liquids, but these fluids should be measured at relatively low flowrates to minimize meter wear.

Applications – Both Liquid and Gas

Coriolis flowmeters (Figure 4-1) are used on both liquid and gas. While they are highly accurate, they have a relatively high initial cost, although some models are now available in the $3,000 range. Despite their higher purchase cost, maintenance costs are normally low.

Figure 4-1. Endress+Hauser Promass X *(Courtesy of Flow Research)*

In the past, nearly all Coriolis flowmeter line sizes (pipe diameters) were limited to six inches or less. The only company with Coriolis flowmeters having larger line sizes was Rheonik (which was purchased by GE Sensing in January 2008). In the past several years, more companies have started manufacturing large size Coriolis flowmeters, including flowmeters for line sizes from 8 to 16 inches. This large-size market is driven by the need for greater reliability and accuracy in liquid and gas flow measurement.

Magnetic Flowmeters Detect the Flow

Magnetic flowmeters use Faraday's Law of Electromagnetic Induction. According to this principle, a voltage is generated in a conductive medium when it passes through a magnetic field. This voltage is directly proportional to the density of the magnetic field, the length of the conductor and the velocity of the conductive medium. In Faraday's Law, these three values are multiplied together, along with a constant, to yield the magnitude of the voltage.

Figure 4-2. Magnetic flowmeter *(Courtesy of KROHNE)*

Magnetic flowmeters (Figure 4-2) use wire coils mounted onto or outside of a pipe. A voltage is then applied to these coils, generating a magnetic field inside the pipe section. As the conductive liquid passes through the pipe, a voltage is generated and detected by electrodes, which are mounted on either side of the pipe. The flowmeter uses this value to compute the flowrate.

Magnetic flowmeters are used to measure the flow of conductive liquids and slurries, including paper pulp slurries and black liquor. Their main limitation is that they cannot measure hydrocarbons (which are nonconductive), and hence are not widely used in the petroleum industry. Magmeters, as they are often called, are quite accurate, though not as accurate as Coriolis and ultrasonic flowmeters, and do not create pressure drop. Their initial purchase cost is relatively high, though most magmeters are priced lower than equivalent Coriolis meters.

Paradigm Case Application

The paradigm case application for magnetic flowmeters involves **conductive fluids that do not contain materials that damage the liner or coat the electrodes and are flowing through a full pipe**. Another limitation on the use of magnetic flowmeters is that they do not work with gases or steam. Because they compute flowrate based on velocity times area, accurate readings require that the pipe be full. In addition, electrode coating and liner damage can degrade the accuracy of magnetic flowmeters.

Applications – the Liquid Leader

Magnetic flowmeters are the leading type of flowmeter for liquid flow measurement, with close to 30 percent of magnetic flowmeter applications in the water & wastewater industry. Other major applications for magmeters are in the food & beverage, chemical, pulp & paper, and metals & mining industries. With relatively high accuracy, magnetic flowmeters are used for billing and even custody transfer applications. Common applications for magmeters are measuring water flow, measuring the flow of water-based chemicals and slurries, process control, sanitary and hygienic applications, and filling machines.

Ultrasonic Flowmeters Time the Flow

Ultrasonic flowmeters were first introduced for industrial use in 1963. There are two main types of ultrasonic flowmeters: transit time and Doppler. Transit time ultrasonic meters have both a sender and a receiver. They send an ultrasonic signal across a pipe at an angle and measure the time it takes for the signal to travel from one side of the pipe to the other. When the ultrasonic signal travels with the flow, it travels faster than when it travels against the flow. The ultrasonic flowmeter determines how long it takes for the signal to cross the pipe in the direction of flow, and then determines how long it takes the signal to cross the pipe against the direction of flow. The difference between these times is proportional to flowrate. Transit time ultrasonic flowmeters are mainly used for clean liquids.

Doppler ultrasonic flowmeters also send an ultrasonic signal across a pipe. However, instead of being sent to a receiver on the other side of the pipe, the signal is reflected off of particles traveling in the flowstream. These particles are traveling at the same speed as the flowstream. As the signal passes through the stream, its frequency shifts in proportion to the mean velocity of the fluid. A receiver detects the reflected signal and measures its frequency. The meter calculates flowrate by comparing the generated and detected frequencies. Doppler ultrasonic flowmeters are used with dirty liquids or slurries.

Ultrasonic flowmeters are used for both liquids and gases. One recent development in the area was the approval by the American Gas Association (AGA) in June 1998 of criteria for using multipath ultrasonic transit time flowmeters for custody transfer of natural gas. This approval gave a major boost to the ultrasonic market in the oil and gas production and transportation industry. Only multipath flowmeters (Figure 4-3) are approved for use in custody transfer.

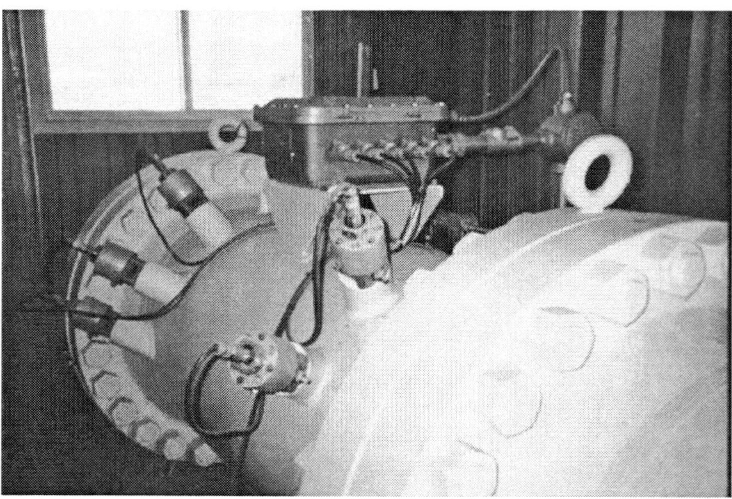

Figure 4-3. Instromet Multipath Ultrasonic Flowmeter ***(Courtesy of Flow Research)***

Most multipath flowmeters use four to six different paths or ultrasonic signals to determine flowrate, with multiple pairs of sending and receiving transducers that alternate in their function as sender and receiver over the same path length. Flowrate is determined by averaging the values given by the different paths. This provides greater accuracy than single path meters.

Paradigm Case Application

The paradigm case application for transit time ultrasonic flowmeters is **clean, swirl-free liquids, and gases of known profile**. The need for high accuracy may require the use of a multipath meter. The most important constraint on ultrasonic flowmeters is that the fluid be clean, although today's transit time meters can tolerate some impurities. A single path ultrasonic meter calculates flowrate based on a single path thorough the pipe, making it susceptible to flow profile aberrations. Multipath flowmeters are more accurate, since they use multiple paths (usually between four and six) to make the flow calculation. Ultrasonic flowmeters can handle liquids and gases, and they can be affected by swirl.

Ultrasonic flowmeters are available in inline and clamp-on models. The paradigm case application for clamp-on models requires taking characteristics of the pipe into account, as well as fluid characteristics. With more than 50 companies worldwide manufacturing ultrasonic flowmeters, a wide variety of models and types are available.

Applications – Rising Star in Custody Transfer

The ultrasonic flowmeter market for custody transfer of natural gas is the single fastest growing niche flowmeter market. Ultrasonic flowmeters are widely used in natural gas transmission and distribution in pipelines to govern the transfer of natural gas from one company to another. Ultrasonic flowmeters are displacing differential pressure and turbine meters for these applications. Advantages of ultrasonic flowmeters include high accuracy for multipath meters, high reliability and virtually no pressure drop.

Other applications for ultrasonic flowmeters include process measurement, check metering, custody transfer of liquids, district heating and monitoring of flare and stack gas emissions. Flare and stack gas monitoring is becoming more important in the United States with the new environmental emphasis from the U.S. government. Here ultrasonic flowmeters compete with multipoint thermal flowmeters and DP flowmeters with averaging Pitot tubes.

Vortex Flowmeters Swirl the Flow

Vortex flowmeters make use of a principle called the von Karman effect. According to this principle, flow will alternately generate vortices in a fluid when passing by a bluff body. In a vortex meter, the bluff body is a piece of material with a broad, flat front that extends into the flowstream at right angles to the flow. Flow velocity is proportional to the frequency of the vortices. Flowrate is calculated by multiplying the cross-sectional area of the pipe times the velocity of the flow.

In some cases, vortex meters require the use of straightening vanes or straight upstream piping to eliminate distorted flow patterns and swirl. Low flowrates also present a problem for vortex meters, because vortex meters generate vortices irregularly under low flow conditions. The accuracy of vortex meters is from medium to high, depending on model and manufacturer.

Vortex flowmeters offer accurate and reliable flow measurement at a competitive price; however, they do not have that one compelling feature that characterizes some other new-technology flowmeters. For example, they do not have the extremely high accuracy of Coriolis flowmeters or multipath ultrasonic flowmeters. Instead, vortex meters offer an accurate and reliable way to measure flow in a wide range of process conditions. Versatility is the hallmark of vortex flowmeters.

Paradigm Case Application

Paradigm case applications for vortex meters are **clean, low viscosity, swirl-free fluids flowing at medium to high speeds**. Because formation of vortices is irregular at low flowrates, ideal conditions for vortex flowmeters include medium to high flowrates. Since swirls can interfere with the accuracy of the reading, the stream should be swirl-free. Any corrosion, erosion or deposits that affect the shape of the bluff body can shift the flowmeter calibration, so vortex meters work best with clean liquids. In addition, vortex meters work best with low viscosity fluids.

Applications – Well Suited for Steam

For many years, vortex flowmeters lacked the industry approvals enjoyed by other meters such as DP, ultrasonic and Coriolis. In January 2007, the Committee on Petroleum Measurement of the American Petroleum Institute (API) approved a draft standard for the use of vortex flowmeters in custody transfer applications. This approval, which applies to liquid, steam and gas applications, has the potential of boosting the vortex flowmeter market. The extent to which this will happen depends on how end users and suppliers respond.

Vortex flowmeters are well suited for steam flow measurement, and they are widely used for that purpose. Steam is the most difficult fluid to measure, and vortex flowmeters are one of the few types of flowmeters that can effectively measure steam flow and handle the accompanying high pressures and temperatures. Measuring steam flow effectively also varies depending on which type of steam is being measured: wet steam, saturated steam, or superheated steam. Steam flowrate is often measured in process plants, and it is also measured for power generation purposes. The other main type of flowmeter used to measure steam flow is differential pressure.

Thermal Flowmeters Heat the Flow

Thermal flowmeters were developed in the late 1960s and early 1970s as an offshoot of research into air velocity profile and turbulence research. This research, conducted by Dr. Jerry Kurz and Dr. John Olin, used hot-wire anemometers. These anemometers consist of a heated thin-film element, but they are too fragile for industrial environments. Kurz and Olin collaborated to form a company called Sierra Instruments in 1973. They developed a thermal flowmeter that was more rugged and hardened than the hot-wire anemometers. When they split up in 1977, Olin stayed with Sierra and Kurz formed Kurz Instruments. Both companies are located in Monterey, California.

Fluid Components International (FCI) took a different approach to developing thermal flowmeters. FCI was founded in 1964 by Mac McQueen and Bob Deane. The company developed flow switches to detect the flow of oil through pipes in the oil patch. Though these thermal switches were not flowmeters, they formed the basis of what later became flowmeters. In 1981, FCI put more sophisticated electronics on its switches, creating thermal flowmeters for gas flow measurement.

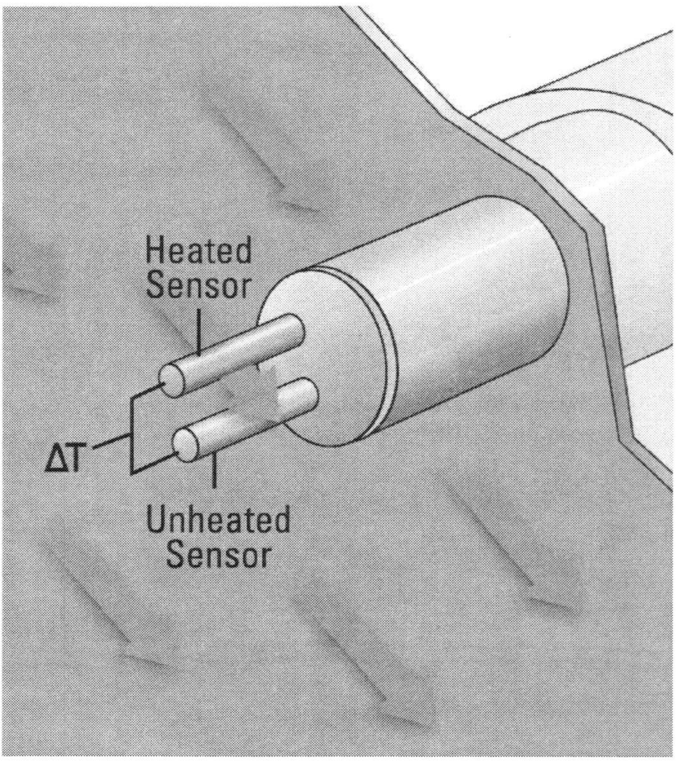

Figure 4-4. Diagram of Thermal Flowmeter *(Courtesy of Fluid Components International)*

Thermal flowmeters (Figure 4-4) work by introducing heat into the flowstream and measuring how much heat dissipates, using one or more temperature sensors. There are two different methods for doing this. One method is called the *constant temperature differential.* Thermal flowmeters that use this method have two temperature sensors. One is a heated sensor and the other sensor measures the temperature of the gas. Mass flowrate is calculated based on how much electrical power is required to maintain a constant temperature difference between the two temperature sensors.

The second method is called the *constant current method.* Under this method, thermal flowmeters also have two sensors, one heated sensor and one that measures the temperature of the flowstream. Power to the heated sensor is kept constant. Mass flow is measured based on the difference between the temperature of the heated sensor and the temperature of the flowstream. Both methods make use of the principle that higher velocity flows result in greater cooling, computing mass flow by measuring the effects of cooling of the heated sensor by the flowstream.

Paradigm Case Application

The paradigm case application for thermal flowmeters is **clean, pure, flowing gases of known heat capacity or clean mixtures of pure gases of known composition and known heat capacity under temperature conditions of 100°C or less**. Thermal flowmeters have limited application to liquids, and they do not work for measuring steam flow.

Applications – CEM Niche

In the 1990s, thermal flowmeters found a niche in handling continuous emissions monitoring (CEM). During this time, new environmental regulations required that companies detect and reduce the emission of nitrogen oxides (NO_X) and sulfur dioxide (SO_2), both of which are principal causes of acid rain. These substances are measured by combining a measurement of flowrate with a measurement of the concentration of NO_X and SO_2.

Thermal flowmeters measure flowrate at a point. Because of the large diameters of the stacks that emit NO_X and SO_2, it is difficult for a single-point flowmeter to accurately measure the flow. In response, suppliers developed multipoint thermal flowmeters that measured the flow at multiple points. Some of these flowmeters have as many is 16 measuring points.

During the early part of the 2000s, environmental awareness was on the back burner in the United States, mainly for political reasons. Beginning in 2009, environmental regulations became more restrictive. In addition, global warming is now widely accepted in scientific and political circles as scientific fact, rather than as speculation. As a result, there is an increasing need to monitor greenhouse gas emissions in a variety of circumstances. Thermal flowmeters are well suited for greenhouse gas monitoring. The following are examples of applications where thermal flowmeters can be used:

- Measurement and recovery of landfill gas
- Ethanol distillation and refining
- Measuring emissions from boilers, steam generators and process heaters
- Recovery of methane from coal mines
- Biomass gasification
- Monitoring of flue gas
- Measurement and monitoring of flare gas flow

Traditional Technology Flowmeters Trace Their Roots to the Mid-1800s

Just as flowmeters that incorporate new technologies are classified as new-technology flowmeters, so flowmeters that incorporate more traditional technologies are classified as traditional technology flowmeters. As a group, these meters have been around longer than the new-technology meters. They generally have higher maintenance requirements than new-technology flowmeters, and while suppliers continue to bring out enhanced traditional technology flowmeters, new-technology flowmeter suppliers are more active in bringing out new products.

Included in the traditional technology category are differential pressure (DP), turbine, positive displacement, open channel and variable area meters. The history of DP meters goes back to the early 1900s, while the beginnings of the turbine meter go back to at least the mid-1800s.

Many of the problems inherent in DP meters are related to the primary elements that they use to measure flow. For example, orifice plates are subject to wear, and can also be knocked out of position by impurities in the flow stream. Turbine and positive displacement meters have moving parts that are subject to wear. The accuracy levels of open channel and variable area meters are significantly lower than those of new-technology flowmeters.

Differential Pressure Flowmeters Constrict the Flow

The groundwork for differential pressure (DP) flow measurement was laid by Bendetto Castelli and Evangelista Torricelli with their work on pressure, and with the idea that flowrate equals flow velocity times the cross-sectional area of the pipe. (Torricelli is most known for his work in developing the barometer.) In 1738, Daniel Bernoulli developed the hydraulic equation for flowrate that is described in Chapter Three. This equation, in various formulations, still forms the basis for DP flow measurement today.

Differential pressure flowmeters require the presence of a primary element to place a constriction in the flowstream. The most common primary elements are:

- Orifice plates
- Venturi tubes
- Pitot tubes
- Flow nozzles
- Wedge elements
- Other

These primary elements are described in Chapter Three, including their historical development and function in flow measurement.

One important way that DP flowmeters are different from other types is that the primary element that is used with a DP flow transmitter to measure flow is often sold separately from the DP flow transmitter. The primary element is often purchased from a company that specializes in primary elements, rather than from a supplier of DP flow transmitters. Some companies that mainly sell primary elements have supplier agreements with multiple DP flow transmitter companies, and supply whatever brand DP flow transmitter a customer specifies.

Some companies have brought out products that include a DP flow transmitter mounted together with a primary element. Emerson Process Management's Rosemount Annubar Flowmeter is an example. These integrated flowmeters make it unnecessary for customers to purchase a primary element separately. They also are calibrated at the factory with the primary element connected to the DP flow transmitter, which is an advantage to the customer.

Types of Primary Elements

There are five main types of primary elements. Which type is best for a specific application depends on a number of factors, including type of fluid, flow velocity, temperature, pressure, viscosity, and other considerations.

Orifice Plates

Orifice plates are the most commonly used type of primary element, and make up more than half the total number of primary elements used to measure flow. Orifice plates are simple in construction and inexpensive to make, and are the most studied and best understood type of primary element. They can be used to measure the flow of liquid, gas, and steam. They do have the disadvantage of being subject to wear, and they can also cause significant pressure loss. Figure 4-5 shows an orifice fitting installed in an oil well gathering station.

Figure 4-5. An orifice fitting installation near Traverse City, Michigan *(Courtesy of Flow Research)*

Venturi Tubes

Venturi tubes are recommended for use with clean liquids, gas, or steam. They can also be used on slurries and dirty liquids. The amount of permanent pressure loss they cause is relatively small, due to their wide opening. Venturi tubes work best with high-speed flows. They have the disadvantage of being relatively expensive, and they can be difficult to install.

Pitot Tubes

Pitot tubes are designed for use with clean liquids, gas, or steam. They do not work as well with low velocity gases and viscous liquids. They can clog when they are used with dirty liquids or gases. Pitot tubes are used with DP flow transmitters to measure air and gas flow in large pipes. They are relatively low in cost and cause little pressure drop. Primary applications for the use of Pitot tubes in large pipe flow measurement include:

- Stack gas flow measurement
- Boiler inlet air flow measurement
- Process gas flow measurement

Flow Nozzles

Flow nozzles are designed for use with clean and dirty liquids, clean gases and saturated steam. They work well with high velocity and high temperature applications. Clean fluids are preferred since removing flow nozzles for cleaning can be difficult. They have a relatively low pressure drop, especially when compared to orifice plates. Flow nozzles are expensive, and are not as well known and studied as orifice plates.

Wedge Elements

Wedge elements are functionally similar to orifice plates, but are designed for flow measurement of liquids containing solids, erosive and abrasive fluids, and high viscosity fluids. The installation requirements for wedge elements are similar to those for orifice plates. Wedge elements typically create less pressure loss than either orifice plates or flow nozzles.

Other Primary Elements

Other primary elements include low loss flow tubes, Dall tubes, laminar flow elements, and the V-Cone element. Low loss flow tubes are designed for reduced pressure loss. Many of them are proprietary, and their features vary with the manufacturer. The Dall tube is a shortened Venturi tube that generates higher differential pressure but lower pressure loss than a Venturi tube. Laminar flow elements are mainly used to measure low flows. The V-Cone is a proprietary element manufactured by McCrometer and designed to operate effectively with reduced upstream piping requirements. Some other companies, including Cameron, now also manufacture cone-style primary elements.

Industry Approvals

Possibly the most famous industry standard for flowrate measurement is AGA-3, which was first published by the American Gas Association in 1955. This report was called "Orifice Metering of Natural Gas, AGA Report #3." It followed two earlier reports: AGA Report #1, published in 1930, and AGA Report #2, published in 1935. Both of these earlier reports covered the use of orifice flowmeters for natural gas flow measurement. In 1975, the American Petroleum Institute (API) adopted AGA-3 as a standard, and submitted it to the American National Standards Institute (ANSI). In 1977, it was adopted as an ANSI/API standard. Since 1977, AGA-3 has been updated several times, most recently in 1992 and 2000.

Product Developments

DP flowmeters have the largest installed base of any type of flowmeter. This is mainly due to the length of time they have been used, the extent to which they have been studied and tested, their relatively low cost, and their relatively straightforward installation requirements. Suppliers have also been working to improve the stability and accuracy of DP flow transmitters and to introduce some improvements to primary elements.

One important product development in this area is the introduction of multivariable DP transmitters. Multivariable DP transmitters measure two or more process variables, usually pressure and temperature, along with differential pressure. Multivariable DP transmitters were first introduced by Bristol Babcock in 1992.

Multivariable DP transmitters are primarily designed to measure mass flow. Their main use is for mass flow measurement of gas and steam. They are becoming increasingly popular as customers seek to reduce costs but also do more measuring of gas and steam flows. While multivariable DP transmitters used to measure flow are typically not as accurate as Coriolis flowmeters, they offer reduced cost and are often compatible with existing flow instrumentation.

Paradigm Case Application

The paradigm case application for differential pressure flowmeters is **clean flowing liquids, gases and steam when pressure loss is not a critical factor**. Different primary elements are better suited to different flow conditions; for example, Venturi tubes do well for clean and dirty liquid flows and for flow measurement in large diameter line sizes. Flow nozzles are designed for high temperature and high velocity applications.

Applications – Widely Used

DP flowmeters are widely used in the oil & gas industry. When fluid is pumped out of an oil well, it typically goes to a separator, which separates the natural gas from the oil and water. DP flowmeters are used to measure the amount of natural gas that comes out of the separator. Another main application of DP flowmeters is for custody transfer of natural gas, especially during natural gas transportation and distribution. Here they compete with turbine and ultrasonic flowmeters. DP flowmeters are also used for gas and liquid flow measurement in subsea applications, where temperature and pressure conditions are extreme.

In addition to their use in the oil & gas industry, they are widely used in refineries and chemical plants to measure flows of liquids, gas and steam. Other industries where they are used include food & beverage, pharmaceutical, pulp & paper, metals & mining, and water & wastewater.

Positive Displacement Flowmeters Trap the Flow

The origins of positive displacement (PD) flowmeters go back to 1843, when Thomas Glover invented the first PD flowmeter. Glover created this meter in response to difficulties with liquid-sealed drum meters, which were invented in the early 1800s.

Glover's meter had sheet metal enclosures with sheepskin diaphragms. Positive displacement flowmeters today (Figure 4-6) are made from cast aluminum and have cloth or synthetic rubber diaphragms.

Figure 4-6. A Positive Displacement Flowmeter ***(Courtesy of FMC Technologies)***

Positive displacement flowmeters work by repeatedly filling and emptying compartments whose volume is known, and then counting the number of times this is done. Flowrate is calculated based on the number of times these compartments are filled and emptied over a known period of time. Positive displacement meters are used to measure both liquids and gases, but they are not used to measure steam.

Since gases change volume with temperature and pressure changes, positive displacement flowmeters used for gases contain sensors to determine gas pressure and temperature. When the pressure and temperature vary significantly from standard conditions, a compressibility factor is applied to the measured volumes.

There are at least six different types of positive displacement meters. These types differ in the way they trap the fluid into compartments of different shape. The following are the six main types of positive displacement flowmeters:

Diaphragm meters are used for gas applications. They have both inlet and outlet diaphragms that capture the gas as it passes through the meter. Differential pressure across the meter causes one diaphragm to expand and one to contract. A rotating crank mechanism helps produce a smooth flow of gas through the meter. This mechanism is connected via gearing to the index, which registers the amount of gas that passes through the meter.

Rotary meters are also used for gas applications, where one or more rotors are used to trap the gas. With each rotation of the rotors, a specific amount of gas is captured. Flowrate is proportional to the rotational velocity of the rotors.

Oval gear flowmeters have two oval gears or rotors mounted inside a cylinder. As the fluid flows through the cylinder, the pressure of the fluid causes the rotors to rotate. As flowrate increases, so does the rotational speed of the rotors. Flowrate is calculated based on this rotational speed. A similar type of gear meter is the **spur gear** meter.

Helical gear flowmeters get their name from the shape of their gears or rotors. These rotors resemble the shape of a helix, which is a spiral-shaped structure. As the fluid flows through the meter, it enters the compartments in the rotors, causing the rotors to rotate. Flowrate is calculated based on the speed of rotation of the rotors.

Nutating disc flowmeters get their name from the idea of nutation, which means nodding or rocking. A nutating disc meter has a round disc that is located inside a cylindrical chamber. The disc is mounted on a spindle. By tracking the movements of the spindle, the flowmeter determines the number of times the disc traps and empties fluid. This information is used to determine flowrate. Nutating disc meters are widely used in residential applications to measure water use in houses. They are also used to measure water use in commercial applications.

Oscillating piston flowmeters have a piston that rotates inside a cylindrical chamber. A control roller guides the piston in its rotation around the cylinder. The piston has holes in it so that the fluid can flow on either side of the piston. As the piston rotates, a specific amount of fluid is trapped. Flowrate is proportional to the piston's rotational velocity.

Paradigm Case Application

Positive displacement (PD) flowmeters work best with clean, non-corrosive and non-erosive liquids and gases, although some models will tolerate impurities. Positive displacement meters also are very good for measuring medium to highly viscous liquids and for measuring liquids and gases at a very low flowrate. Due to their high accuracy, PD meters are used at residences to measure the amount of gas or water used. They are also used commercially to measure gas and water flow at many businesses.

Some designs require that only a lubricating fluid be measured, because the rotors are exposed to the fluid. PD meters differ from turbine meters in that they handle medium and high viscosity liquids well. For this reason, they are widely used to measure the flow of hydraulic fluids. PD meters above ten inches in diameter tend to be heavy, large, and relatively expensive. PD meters require very little upstream piping, and can easily handle low flows. Pressure drop can be an issue.

Applications – Four Distinct Areas

The positive displacement market is divided into four application areas:

- Water
- Oil
- Industrial liquids
- Gas

PD meters for water applications are mainly focused on utility markets. This includes residential, commercial and industrial applications. PD meters are widely used in homes for measuring how much water is used. PD meters for oil applications are widely used in trucking terminals for loading hydrocarbons into trucks and for offloading them from trucks. They are also used in aviation applications and for in-plant measurement within refineries and chemical plants. PD meters are also used to measure the flow of other industrial liquids that are not hydrocarbon-based in a variety of process plants.

PD flowmeters for gas applications are widely used in utility applications to measure the consumption of gas at commercial buildings. Typically, line sizes for these applications are from 1½ inches to 10 inches. Diaphragm meters are well-entrenched in this market, although there is a shift occurring from diaphragm to rotary meters. PD meters also compete with turbine meters for gas applications, but turbine meters are mainly used for higher speed flows in larger line sizes.

Turbine Flowmeters Spin with the Flow

The history of turbine flowmeters goes back to the late 1700s. Richard Woltman is generally credited with being the inventor of the first turbine flowmeter in 1790. Woltman was a German engineer who studied hydraulic engineering. He dedicated his life to the Department of Ports and Navigable Waters of Hanover. He wrote several books about hydraulic engineering, and formulated equations for calculating the loss of energy in pipes and canals.

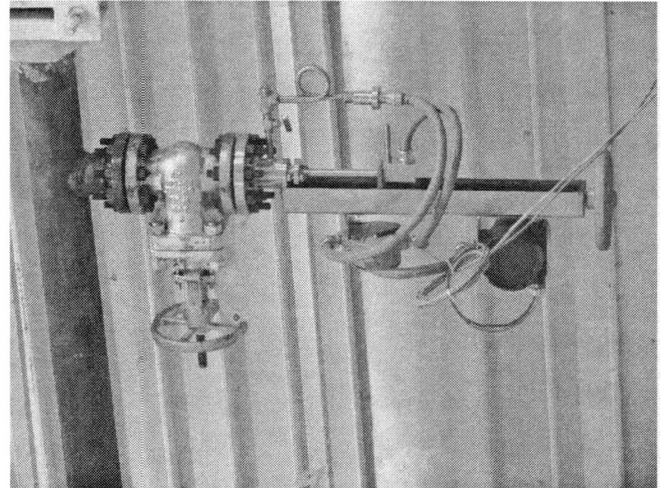

Figure 4-7. An Insertion Turbine Flowmeter *(Courtesy of Spirax Sarco)*

The word *turbine* is derived from a Latin word that means "spinning thing." Turbine flowmeters have a spinning rotor with propeller-like blades. The rotor is mounted on bearings in a housing, and it spins as water or another fluid passes over it. Flowrate is proportional to the rotational speed of the rotor. A number of different methods are used to detect rotor speed, including a mechanical shaft and an electronic sensor.

Turbine meters differ according to the design of the spinning rotor. There are six main types of turbine flowmeters, described as follows:

Axial turbine meters have a rotor that revolves around the axis of flow. Most flowmeters for oil measurement and for measuring industrial liquids and gases are axial flowmeters. Axial meters differ according to the number of blades and the shape of the rotors. Axial meters for liquids have a different design from axial meters for gas applications.

Jet turbine meters are primarily used for municipal water measurement, although some are also used for industrial water measurement. Jet meters are of two types:

- Single jet
- Multi-jet

With single and multi-jet meters, the flow of water impinges on a rotor. Multi-jet meters have orifices that the water is forced through. As water passes through the orifices, a stream of water or "jet" is formed. These jets impact the impeller blades, causing them to revolve.

Paddlewheel turbine meters have a shaft at right angles to the flowstream, and the shaft and bearings are outside of the flowstream. They have a lightweight paddlewheel that spins in proportion to flowrate, and are used for measurement of low-speed flows.

Pelton wheel turbine meters work somewhat like paddlewheel meters, but have a single size rotor with straight blades that spin in proportion to flowrate. Pelton wheel meters are based on the design of the Pelton water wheel.

Propeller turbine meters are bulk meters used mainly to handle dirty liquids. Propeller meters have helical-shaped blades that are longer than the blades of most other turbine meters. They also have fewer blades than the rotors of most other turbine meters.

Woltman turbine meters have a rotor whose axis is in line with the direction of flow. Woltman meters are water meters used for larger volume applications. They are sometimes called "bulk" meters. Woltman meters get their name from Reinhard Woltman, who is generally credited with inventing the first turbine meter in 1790.

Compound meters are difficult to classify because they incorporate two meter technologies. Compound meters are designed to handle both high flowrates and low flowrates. They are often installed in buildings such as office and apartment buildings that have short periods of high flowrates and also extended periods of low flowrates. Many compound meters incorporate both turbine and positive displacement flow technology. The turbine component handles the high flowrates and the positive displacement component handles the low flowrates. Some compound meters have jet-type technology for low flowrates.

Paradigm Case Application

Turbine meters are used to measure the flow of both liquids and gases. Paradigm case conditions for turbine flowmeters include **clean gases or clean low-viscosity liquids flowing at medium to high speeds**. Dirt or impurities in the liquid or gas can damage the meter. Turbine meters are also sensitive to viscosity. A low viscosity fluid is best for a turbine meter. Straight run prior to the meter is recommended, since turbine meters are sensitive to swirl and to flow profile effects. Turbine meters for liquid and gas require different designs, due to the different densities involved.

Applications – Widely Used for Commercial Applications

Like positive displacement meters, turbine meter applications are divided into four application areas as follows:

- Water
- Oil
- Industrial liquids
- Gas

While turbine meters are not so widely used for residential applications, they are widely used for commercial applications, especially for larger buildings. Positive displacement meters for residential applications for liquids are limited to 1½ inch and 2-inch line sizes. Turbine meters for commercial applications, by contrast, start at 1½ inches and go up to 8, 10, 12 or even 16 inches. While there is some overlap with positive displacement meters at the low end in terms of line size, most turbine meters are used for line sizes that are larger than positive displacement meters can accommodate. This is true for turbine meters used for commercial utility applications. Positive displacement meters for industrial applications come in larger line sizes.

Turbine meters are widely used in trucking terminals for loading and unloading hydrocarbons onto and from trucks and for aviation applications. Turbine meters face challenges from new-technology meters in oil and industrial liquids flow measurement. For industrial oil measurement, the main challenges are from ultrasonic and Coriolis meters.

Turbine flowmeters are widely used for the custody transfer of natural gas. Here they compete mainly with ultrasonic and differential pressure flowmeters. Ultrasonic flowmeters are also widely used for custody transfer gas flow measurement, especially for larger size pipes (12 inches and above). In 1998, the American Gas Association (AGA) issued its AGA-9 report, outlining criteria for using ultrasonic flowmeters for custody transfer of natural gas. This has resulted in a major boost for ultrasonic flowmeters, mainly at the expense of turbine and differential pressure meters.

Open Channel Flowmeters Guide the Flow

Open channel flow refers to flow in rivers, streams, canals, aqueducts, irrigation ditches and partially filled pipes. While most flow measurement occurs in closed pipes, open channel flows occur either in partially filled pipes or outside of pipes altogether.

The history of open channels goes back to Roman times. The Romans built a network of aqueducts designed to supply water to their cities. In modern times, Marquis Giovanni Poleni developed the first equation for calculating flowrate over a weir in the 18th century. Poleni was an Italian astronomer, physicist and mathematician. This was at about the same time that Henri Pitot invented what is today called the Pitot tube. In the early 19th century, another Italian named Giorgio Bidone, who taught at the University of Turin, investigated flowrate over a weir. Julius Weisbach further analyzed flow over a weir and in 1845 developed an equation that is similar to the one used today.

One way to distinguish between open channel flow and closed pipe flow is to think of it as the difference between gravity-induced (or gravitational) flow and pressurized flow. Flow in open and uncovered channels such as streams depends on gravity. Flow in partially filled closed conduits, such as drain pipes and culverts, also depends on gravity. By contrast, flow in closed pipes that are completely full occurs under pressure.

Methods Used to Measure Open Channel Flow

A number of methods are used to measure flow in open channels. Many different factors help determine which method is best for a given application. These factors include the nature of the application, budget, accuracy needs and reporting requirements. This section discusses different methods for measuring open channel flow.

Use of Weirs and Flumes: A common method of open channel flow involves the use of a hydraulic structure such as a weir or flume. These hydraulic structures are called primary devices. A primary device is a restriction that is placed in an open channel and that has a known depth-to-flowrate relationship. Once a weir or flume is installed, a measurement of the depth of the water is used to calculate flowrate. Charts are available that correlate various water depths with flowrates, taking into account different types and sizes of weirs and flumes.

In some respects, a weir or flume in an open channel operates like an orifice plate used to measure flowrate in a closed pipe. An orifice plate is a restriction placed in a pipe that causes a pressure drop across the plate. By measuring the pressure differential across the orifice plate, flowrate can be determined, using an equation. The difference is that for open channel flow, water level is the variable measured instead of difference in pressure.

A weir resembles a dam placed across an open channel. It is positioned in such a way that the liquid can flow over it. Weirs are classified according to the shapes of their openings. Types of weirs include:

- V-Notch
- Rectangular
- Trapezoidal

Water depth is measured at a specific place upstream from the weir. An equation for determining flowrate is associated with each type of weir.

A flume is a specially shaped portion of the open channel with an area or slope that is different from the channel's slope or area. The velocity of the liquid increases and its level rises at it passes through the flume. Types of flumes include:

- Parshall
- Palmer-Bowlus
- Leopold-Lagco
- H-type
- Trapezoidal
- Cutthroat

To determine flowrate, liquid depth is measured at specified points in the flume. An equation is associated with each kind of flume, taking flume size into account.

Flow measurement with weirs and flumes also requires a secondary device to measure level. There are a number of technologies used to measure level. These include:

- Submerged pressure transducer
- Float
- Ultrasonic
- Bubbler
- Electrical

Area Velocity: Flow can be measured without a hydraulic structure such as a weir or flume. In the area velocity method, the mean velocity (V) of the flow is calculated at a cross-section, and this value is multiplied by the flow area (A). Normally, this method requires that two measurements be made: one to determine mean velocity, and another to determine depth of flow. Flowrate (Q) is determined according to the continuity equation:

$$Q = V \times A$$

The area velocity method is used when it is not practical to use a weir or flume and for temporary flow measurements. Examples include influx and infiltration studies and sewer flow monitoring.

The velocity measurement is made using a variety of technologies, including:

- Doppler
- Transit time
- Electromagnetic
- Radar

Dilution: Using the dilution method, a tracer solution is added to the flowing water. Somewhere downstream, a measurement is made of the amount of solution contained in the flowing water. This amount is used to calculate flowrate, based on a theoretical formula. Radioactive and fluorescent dyes are examples of tracers used in the dilution method. Two techniques used in dilution flow measurement are the constant rate injection method and the total recovery method.

Timed-Gravimetric: In the timed-gravimetric method, the liquid is captured in a container for a specified period of time. The liquid is then weighed, using a force measuring device such as a beam scale or load cell. The timed-gravimetric method gives high accuracy, but is not suited to continuous flow measurement.

Manning Formula: Another method of flow measurement involves using a modification of a formula that Robert Manning, an Italian civil engineer, first proposed in 1889. His original formula was modified in the 1930s. It is used to calculate flowrate based on values such as cross-sectional area of flow, roughness of the conduit and slope of the water surface. Use of this formula does not require the presence of a primary device. However, it is less accurate than the area velocity method because flow velocity is calculated based on assumed values rather than being measured. Another name for the Manning formula is the slope-hydraulic radius method.

Paradigm Case Application

The paradigm case for open channel flowmeters depends on the type. **For weirs, the paradigm case application is a clean free-flowing liquid stream with sufficient slope for relatively rapid flow across the weir**. Weirs do not work well on channels with a flat slope, or when pressure loss is a consideration. The liquid should be clean so that sand, silt or other solid materials do not collect behind the weir. This might interfere with proper measurement.

The paradigm case for flumes is for **a free-flowing liquid stream flowing at a relatively high velocity**. Flumes work well when the slope is relatively flat and pressure loss is a consideration. Flumes can tolerate some impurities since they will be swept through the flume, and will not catch on the edge like they might on a weir.

The paradigm case for area-velocity flow measurement is **measurement of liquid flow in partially filled pipes with diameters of six inches and larger**. The area-velocity method works better for partially filled pipes than for open channel streams because it is difficult to get an accurate measure of area without the hydraulic structure of a pipe, and the bottom of a stream can be irregular. The area-velocity method works well for temporary flow-monitoring applications such as infiltration and inflow studies, since it does not require the installation of a weir or flume.

Applications – Exposed Liquids

Open channel flowmeters measure the flow of liquids that are exposed to the atmosphere. They are widely used in the water & wastewater industry to measure water and sewage flows, including flows in storm and sanitary sewer systems, sewage treatment plants and irrigation systems. They are also used to measure effluent in the mining, pulp & paper, refining, chemical and power industries.

Variable Area Flowmeters Float the Flow

Karl Kueppers invented the first variable area flowmeter, which had a rotating float, in Aachen, Germany in 1908. In the following year, Felix Meyer founded the company Deutsche Rotawerke GmbH in Aachen, the forerunner of the company that is today known as Rota Yokogawa. KROHNE started producing variable area flowmeters in Duisburg, Germany in 1921. Since that time, many other companies have begun manufacturing these flowmeters. Variable area flowmeters are also called *rotameters*, though Rotameter is a registered trademark.

Most variable area (VA) flowmeters consist of a tapered vertical tube that contains a float. The upward force of the flowing fluid is counterbalanced by the force of gravity. The point at which the float position stays constant indicates the volumetric flowrate, which can be often read on a scale on the meter tube. VA meter tubes are made of metal, glass or plastic. Metal tubes are the most expensive type, while the plastic tubes are lower in cost. Metal tubes are used for high-pressure applications.

While most VA meters can be read manually, some also contain transmitters that generate an output signal that can be sent to a controller or recorder. While VA meters should not be selected when high accuracy is a requirement, they do very well when a visual indication of flow is sufficient. They are effective at measuring low flowrates and can also serve as flow/no-flow indicators. VA meters do not require electric power and can safely be used in flammable environments.

One important development for variable area flowmeters is the development of meters with a transmitter output. The HART protocol is available on some meters. This turns the VA meter into more than a visual indicator, and makes it possible to use the meter to control and to record. A class of VA meters called purgemeters have been developed to handle a variety of low flow applications. Areas of research in VA meters include float design and materials of tube construction, especially metals.

Paradigm Case Application

Paradigm case applications for variable area flowmeters are **clean flowing liquids of low viscosity, where high accuracy is not a requirement**. VA meters are good for making spot checks of flowrates and for determining flow/no-flow situations.

Applications – Low Accuracy

The use of variable area flowmeters in process applications is limited by their low accuracy. In addition, since many VA meters have to be read manually and do not produce an output signal, they are not widely used for monitoring and control applications. VA meters can measure gas flow, provided the gas is dense enough and flowing at a high enough velocity to lift the float. They are not used for steam flow measurement.

Emerging Technology Flowmeters Enter the Scene in the 21st Century

This section discusses two emerging technologies for flow measurement:

- Sonar
- Optical

Sonar Hears the Flow

SOund Navigation And Ranging (SONAR) flow technology provides a non-intrusive and robust way to accurately and repeatedly measure a wide range of singlephase and multiphase flows, including pulp slurry, sewage and flows with entrained air.

This emerging technology is based on sonar techniques used for decades in submarine tracking and other underwater acoustic applications. CiDRA, which introduced its SONARtrack™ flowmeters in October 2003, developed it for the highly demanding oil & gas production industry. CiDRA is one of two manufacturers offering sonar flow processing; the other is Expro Meters, Inc.

Sonar flow processing captures two measurements: 1) the velocity of naturally occurring turbulent eddies within the process flow and 2) the speed at which sound waves propagate through the fluid within the pipe. Sonar flowmeters use both these values to compute volumetric flowrate.

Unlike some conventional flowmeter technologies (e.g., differential pressure, turbine) that create pressure drops by introducing obstructions, sonar measurement measures naturally occurring turbulence, eliminating pressure loss and clogging.

Some applications for sonar include:

- Oil & gas
- Oil sands processing
- Mineral processing
- Chemical
- Paper & pulp
- Consumer products
- Water and wastewater treatment

- Power generation
- Pharmaceuticals
- Food & beverage industry

Optical Sees the Flow

Optical flow metering is an emerging technology for natural gas and flare measurement. It uses optical laser technology to measure the actual speed of particles rather than the speed of sound through a medium (as in ultrasonic flowmeters) and conductivity (as in magnetic flowmeters).

Commercial instruments are essentially optical transit time velocimeters, which rely on optical scintillation technology. They measure the speed of particles that accompany natural and industrial gases flowing through a flare pipe. As they pass through a laser beam, these particles cause the laser beam to scatter and generate a pulse signal. The speed of the particles is determined by measuring the time between signals based on the known distance between two photodetectors.

Optical flowmeter (OFM) companies claim their technology offers unmatched effectiveness for flare flowrate measurement. This is good news for the environment and the oil & gas industry, which sees potential profits go out the pipe every year. The amount of flared gas in 2006 was valued around $40 billion, up 14% from 2004, according to a World Bank-financed study. Flaring has dropped significantly in Norway and Canada, which have developed flare metering initiatives.

Photon Control, a leader in the field, is also developing a line of optical flowmeters to measure flowrates of gases and clear liquids in process control and custody transfer applications. Optical Scientific Inc. is also a leader in optical flow measurement.

Paradigm Case Selection Method

While various selection methods have been devised, this chapter presents a step-by-step method that begins by matching the application under study with the **paradigm case applications** for various types of flowmeters. It then advocates looking at application, performance, cost and supplier criteria in order to select a flowmeter. A statement of this paradigm case method follows:

1. Every type of flowmeter is based on a physical principle that correlates flowrate with some set of conditions. This principle determines the paradigm case application for this type of flowmeter. When selecting a flowmeter, begin by selecting the types of flowmeters whose **paradigm case applications** are close to your own.

2. Make a list of **application criteria** that relate to the flow measurement you wish to make. These conditions may include type of fluid (liquid, steam, gas, slurry), type of measurement (volumetric or mass flow), pipe size, process pressure, process temperature, condition of fluid (clean or dirty), flow profile considerations, fluid viscosity, fluid density, degree of turbulence, range and others. From those types of flowmeters selected in Step 1, select those that meet these application criteria.

3. Make a list of **performance criteria** that apply to the flowmeter you wish to select. These include reliability, accuracy, repeatability, range and others. From those types of flowmeters selected in Step 2, select the ones that meet these performance criteria.

4. Make a list of **cost criteria** that apply to your flowmeter selection. These include initial cost, cost of ownership, installation cost, maintenance cost, and others. From the types of flowmeters chosen in Step 3, select the types that meet your cost conditions.

5. Make a list of **supplier criteria** that govern your selection of a flowmeter supplier. These include types of flowmeters offered, company location, service, responsiveness, training, internal requirements and others. From the types of flowmeters listed in Step 4, select the suppliers that meet your criteria.

6. For the final step, review the meters that are left as a result of Step 4 and the suppliers listed as a result of Step 5. Review the application, performance and cost conditions for the remaining flowmeter types and select the one that best meets all these conditions. Now **select the best supplier for this flowmeter** from those suppliers listed as a result of Step 5.

Surveys of flowmeter users consistently show that reliability and accuracy are the two performance criteria rated highest in importance by users when selecting flowmeters. Among new-technology flowmeters, Coriolis flowmeters provide the highest accuracy, followed by ultrasonic and magnetic meters. In terms of cost, many users are now distinguishing between purchase cost and total cost of ownership. As a result, they may be willing to pay more for a flowmeter if it promises reduced maintenance costs.

How are decisions actually made in a plant about which flowmeter to buy? Users often choose to replace like with like. There are several reasons for this. Inventories of parts and supplies are often built up in a plant based on a particular type of flowmeter. It can be very expensive to train personnel to install, use and maintain a new type of flowmeter. Changing flowmeter types sometimes means changing flowmeter suppliers, which can be difficult.

The above reasons help explain why differential pressure flowmeters still have the largest installed base of any flowmeter type. The battle for the hearts and minds of users is largely between the suppliers of new-technology flowmeters and the suppliers of differential pressure flowmeters. It is less a battle among the suppliers of new-technology flowmeters, although new lower-cost Coriolis flowmeters may begin to impinge on the magnetic flowmeter market.

Multivariable flowmeters represent one way the differential pressure flowmeter suppliers are responding to the challenge of new-technology flowmeters. Multivariable flowmeters usually measure pressure and temperature in addition to flow.

Multivariable vortex and multivariable magnetic flowmeters have also been developed, and it is likely that more types of multivariable flowmeters will be introduced in the future. This ongoing drama is definitely worth watching.

Refer to Table 3-1 for a comparison of the advantages and disadvantages of different types of DP flowmeter primary elements. Table 4-1 gives the advantages and disadvantages of different types of new-technology flowmeters. Table 4-2 gives the principles of operation of new-technology and DP flowmeters.

Table 4-1. Advantages and Disadvantages of New-Technology Flowmeters

Flowmeter Type	Advantages	Disadvantages	Liquid, Steam, or Gas	Pipe Size	Comment
Coriolis	High accuracy Measures mass flow directly	Becomes expensive and unwieldy in pipe sizes over four inches High initial cost Sensitive to vibration	Liquid, Gas	1/16 inch to 6+ inches	Measures mass flow directly
Magnetic	Obstruction-less High accuracy No pressure drop	Cannot meter nonconductive fluids (e.g., hydrocarbons) Relatively high initial cost Electrodes subject to coating	Liquid	1/10 to 100+ inches	Limited use in the petroleum industry because it does not meter hydrocarbons
Ultrasonic: Single Path Transit Time	High accuracy Nonintrusive	High initial cost Requires clean fluids Sensitive to swirl	Liquid, Gas	½ inch and up	Used for check metering applications
Ultrasonic: Multipath Transit Time	High accuracy Nonintrusive	High initial cost Requires clean fluids Sensitive to swirl	Liquid, Gas	4 inches to 36 inches	Approved for custody transfer of natural gas
Ultrasonic: Doppler	Operates on dirty liquids Nonintrusive	Low-medium accuracy	Liquid	½ inch and up	Limited accuracy but one of few meters designed for dirty liquids
Vortex	Medium-high accuracy	Vibration can affect accuracy Lacks industry approvals	Liquid, Steam, Gas	½ inch to 16 inches (inline)	Widely used for steam measurement

Table 4-2. New-Technology and DP Flowmeter Principles of Operation

Flowmeter Type	Technology
Coriolis	Fluid is passed through a vibrating tube, which causes the tube to twist. Mass flow is proportional to the amount of twisting by the tube.
Magnetic	A meter that creates a magnetic field within a pipe, typically using electrical coils. As electrically conductive fluid moves through the pipe, it generates a voltage. Flowrate is proportional to amount of voltage, which is detected by electrodes.
Ultrasonic: Single Path Transit Time	A flowmeter that measures the time it takes an ultrasonic pulse or wave to travel from one side of a pipe to the other. This time is proportional to flowrate.
Ultrasonic: Multipath Transit Time	A flowmeter that uses multiple ultrasonic paths to calculate flowrate; typical number of paths is three to six.
Ultrasonic: Doppler	A flowmeter that calculates flowrate based on the shift in frequency observed when ultrasonic waves bounce off particles in the flowstream.
Vortex	A bluff body is placed in a flowstream; as flow passes this bluff body, vortices are generated. The flowmeter counts the number of vortices, and flowrate is proportional to the number of vortices generated.
DP: Orifice Plate	A flat metal plate with an opening in it; a DP transmitter measures pressure drop and calculates flowrate.
DP: Venturi Tube	A flow tube with a tapered inlet and a diverging exit; a DP transmitter measures pressure drop and calculates flowrate.
DP: Pitot Tube	An L-shaped tube inserted into a flowstream that measures impact and static pressure; the opening of the L-shaped tube faces directly into the flowstream. The difference between impact and static pressure is proportional to flowrate.
DP: Averaging Pitot Tube	A Pitot tube having multiple ports to measure impact and static pressure at different points. Flowrate is calculated by DP transmitter based on average of difference in pressure readings at different points.
DP: Flow Nozzle	A flow tube with a smooth entry and sharp exit; flowrate is calculated based on difference between upstream and downstream pressure.

Users Migrate from Traditional Technology to New-Technology Flowmeters

A gradual shift in the flowmeter market is taking place. Users are migrating from traditional technology flowmeters to new-technology flowmeters. This means that, in many cases, users are selecting Coriolis, ultrasonic and magnetic flowmeters instead of differential pressure, positive, displacement or turbine meters. This is a long-term migration that has been going on for at least the past 15 years. There are a number of reasons for this migration:

Accuracy. New-technology flowmeters typically have higher accuracy than traditional meters, and better maintain it over time. Coriolis flowmeters are the most accurate meters made, and ultrasonic flowmeters are also highly accurate – especially the multipath meters. Orifice plates are subject to wear, and the spinning rotor on turbine meters also wears over time. Since new-technology flowmeters do not have moving parts, apart from the vibration that Coriolis meters are subject to, their accuracy is not as likely to degrade over time.

Reliability and reduced maintenance. New-technology flowmeters tend to be more reliable over time than traditional meters. Coriolis, ultrasonic, magnetic, vortex and thermal flowmeters do not have moving parts (apart from the vibration of Coriolis flowmeters). Positive displacement and turbine are mechanical meters with moving parts that are subject to wear. Orifice plates are subject to wear, and they can also be knocked out of position. Many companies have fewer engineers in place, and are looking for flowmeters with minimal upkeep and maintenance.

Industry approvals. For many years, traditional technology flowmeters were the only ones that had approvals from industry associations like the AGA and the API. DP and turbine meters especially benefited from these approvals. Since 1998, ultrasonic and Coriolis flowmeters have been granted approvals as well, enabling their use in custody transfer applications. In January 2007, the API approved vortex meters for custody transfer as well. As a result, many new-technology flowmeters have the same type of approvals enjoyed for so long by traditional meters.

New project selection. It is true that, in some cases, users will choose to replace one type of flowmeter with another of the same type, when replacement is needed.

This trend favors traditional flowmeters, since traditional meters have a larger installed base than new-technology flowmeters. However, for new projects, users are more likely to select new-technology flowmeters in many cases. That is because new-technology flowmeters are the most up-to-date in terms of product features and enhancements. They are more likely to have the latest communication protocols, such as Foundation Fieldbus and Profibus. New-technology flowmeter performance is at a higher level than most traditional meters, and they fit in better in an instrumentation network. For all these reasons, new-technology flowmeters are often favored for new capital construction projects.

Suppliers of traditional flowmeters are responding by improving the quality of their products. Turbine flowmeter suppliers are using ceramic ball bearings for longer life. Multivariable DP flowmeter suppliers are making DP flowmeters capable of measuring mass flow at a lower cost than Coriolis meters. Some variable area flowmeters now even provide an output signal, negating the need for them to be manually read. These are positive product developments for traditional technology flowmeters.

Despite these enhancements to traditional meters, the long-term trend still favors new-technology flowmeters. More research and development is focused on new-technology flowmeters than on traditional meters, and there are more new-technology products and product enhancements released than there are enhancements to traditional meters. While traditional meters still have an advantage in installed base, the advantage for current sales goes to new-technology flowmeters, and this will most likely be true for the foreseeable future.

The Future of Flow Measurement

What lies ahead for flow measurement? It is likely that more new flow technologies will emerge. The current development of sonar and optical flowmeters shows that this is possible. These new technologies are as likely to come from independent entrepreneurs and inventors as from the laboratories of the major suppliers.

Wireless technology will have an impact on the flowmeter market. Wireless technology has advantages in remote and hazardous locations and where portability is desirable. But it does have some drawbacks, such as susceptibility to being monitored by outside, unwanted sources. One area where wireless has been highly successful is in the remote monitoring of electric and gas meters by utility companies.

Some of the distinctions between existing technologies may disappear, as suppliers produce combined flowmeters that incorporate more than one flow technology. Compound flowmeters, which incorporate both positive displacement and turbine meters, are one example. No doubt there are other ways to create "duometers" that incorporate two flow technologies in one. This may be a way to broaden the range of conditions a flowmeter can handle, or even to increase the performance of the meter.

Renewable energy will provide new opportunities for flow measurement. New forms of energy such as wind power, solar and geothermal will require their own forms of flow measurement. These alternative energy technologies will inevitably grow as crude oil prices rise and as environmental considerations require less reliance on carbon-based fuels. This is one area that is wide open for development, and there are many opportunities here for companies that want to take advantage of them.

Exactly how these technologies will shape up is uncertain, but one thing is sure – they will all need to "go with the flow."

Definitions of Key Flow Terms

The following are definitions of some of the key terms used in describing flowmeters and flow measurement.

Accuracy: The accuracy of a flowmeter is its ability to provide a volumetric or mass flow reading that corresponds to the volumetric or mass flowrate that is actually passing through the flowmeter at the time the measurement is made. Accuracy is generally stated in terms of percent and is typically stated as a percentage of how close to the actual flowrate the flowmeter reading is. A flowmeter with a 1.0% accuracy level is accurate to within 1% of the actual reading.

Mass Flowrate: Mass flowrate is a measurement of the mass of fluid passing through a specific region of a pipe in a specified period of time. Mass flowrate is often measured in kilograms per minute (kg/min) or pounds per minute (lbs/min). Mass flowrate varies with changes in temperature and pressure for gases and steam, but changes very little with temperature and pressure changes for liquids.

Precision: Precision in flow measurement refers to the number of digits to which a measurement is made. It is not the same as accuracy since it does not imply a correspondence with the actual value of the flowrate in a pipe or open channel.

Repeatability: The repeatability of a flowmeter is the ability to produce exactly the same measurement under exactly the same conditions. Repeatability is sometimes viewed as being as important as accuracy.

Stability: Accuracy of measurement over time.

Turndown: The turndown of a flowmeter is the ratio of the maximum flowrate that the flowmeter will measure within its accuracy level to the minimum flowrate that can be measured within the same accuracy level.

Velocity: The flow velocity is the speed at which flow passes through a specified point or location within a pipe or open channel. Velocity is often measured in feet per second (feet/second) or in meters per second (meters/second). Velocity × pipe cross-sectional area = volumetric flowrate.

Volumetric Flowrate: Volumetric flowrate is the total volume of fluid that moves within a pipe or open channel in a set period of time. Volumetric flowrate is measured in a variety of units including gallons per minute and cubic meters per minute.

Morley's Point:
Flow

This is one of the more difficult topics to review. My experience is mostly in discrete manufacturing, which does not deal with the concept of flow in most cases. So I decided to look it up. All the definitions have a common characteristic: the continuous movement of mass. The terms used were "continuous progression," "uninterrupted moving," "continuously running." The phrase "interrupted the flow" can be used, but the flow being discussed here is continuous during the time of observation.

I feel that Jesse's tome is excellent – a valuable discourse, indeed. But he examined the measurement technology instead of the system being used for this flow. There is no risk analysis and no systems analysis of the utilization of flowmeters. The technology he described that is used to measure flow was astounding. The engineers of the suppliers should be congratulated but I felt as though I was missing the point of engineered flowmeters. Clearly the measurement of flow is important, but what does it mean to the maintenance, the lifetime or the accuracy of the meter?

I asked some of my system friends in the fluid processing business. They feel the real cost of a flowmeter is the cost of maintenance. The system engineers fit the meter to the application without concern for the system. It takes, according to my interview, two to five hours to replace a four-inch flowmeter. The cost of maintenance is on the order of seven times more than the cost of the meter itself. Vendors typically give life cycle but do not consider the cost of maintenance or the cost of process downtime for maintenance or meter failure. I know that for trucks – when a truck is made every three minutes – being down for a day is a lot of trucks not being made. I assume that the problem is the same for high volume processes such as oil, wastewater and steam.

Specification of flowmeters only includes the pipe size, the required accuracy and the lifetime, depending on the flowmeter chosen. It has to be designed to operate for 10 years with no maintenance. But I seem to meet a brick wall when I ask about the future. Everybody says, "We know flowmeters" without regard to whether they know them or not. The enemy of "better" is "good enough." We have a groove in our brain that suggests that flowmeter technology is current and adequate. That's correct. It is current and it is adequate but that doesn't mean it fully meets modern system requirements.

Let's talk in terms of the future. Every flowmeter will be hooked up to a wireless system and have links to the cloud. The connection of the system topology will be in the cloud and not in the wiring of the system. Every sensing and measurement device (they are not the same thing) will have a common interface to a cloud database. The physical location of the sensing and measurement locations is independent of its connection to other parts of the system (in other words, put the toaster anywhere you want).

The specification of the future will include elements, such as risk analysis, wireless compatibility and availability. I'm not sure that the accuracy everyone quotes is a necessity for flowmeters in all applications. As a casual reader can tell, I seem to have a system bias. Oh well.

"Put your hand on a hot stove for a minute, and it seems like an hour. Sit with a pretty girl for an hour, and it seems like a minute. THAT'S relativity." – Albert Einstein

Chapter Five

Measuring Time as It Flows On

Overview

Albert Einstein

When did time begin, or has it always existed? Does the phrase "the beginning of time" have a reference? Does time exist apart from the measurement of time? Will time ever end? These questions reveal some of the fundamental difficulties in understanding the nature of time. Time is something that seems to be always present and yet is hard to grasp directly. We mainly experience time through the means of measuring it, such as clocks, watches and calendars, rather than directly.

Is time an objective quality like weight or hardness? Time seems to have an objective quality, but this objective quality does not appear to be directly perceivable. A clock does not sense time in the way in which a pressure sensor senses pressure. A pressure sensor senses pressure in a physical body and responds according to a rule that correlates the amount of pressure with some physical state in the sensor. This physical state may then be converted to a pressure reading. A mechanical clock, by contrast, is entirely self-directed, and responds to a physical spring or a predetermined movement. A digital clock has a crystal oscillator or a gearing mechanism and displays the time in the form of numbers. A clock or watch runs parallel with time rather than sensing it. An accurate clock or watch is one that conforms to a standard of time that has been independently established, such as the atomic clock. But the atomic clock itself is not a time sensor; instead, it just serves as a standard of correct time measurement.

Our concept of time is inevitably linked to human language and to the network of concepts by means of which we perceive the world. Would time exist without being measured? In fact, is there really any such thing as time? Albert Einstein, who transformed the way we understand time with his theory of relativity, referred to time as "the fourth dimension" of a space-time continuum. He also maintained, "People like us, who believe in physics, know that the distinction between past, present, and future is only a stubbornly persistent illusion." That notion would seem to negate the idea of fixed time as a reality.

Yet certainly time as conventionally perceived — defined until modern times by the movement of the sun and moon around the earth, and still defined as the progression of events from the past to the future — did exist before the invention of clocks and watches. It even existed before conscious beings, human or animal, began to grace this planet and observe time's passage. Our understanding of the time before recorded time depends, however, on our ability to project our current system of time measurement back on earlier days. Scientists have calculated the age of our universe at about 15 billion years. This could be converted into a number of days, minutes and seconds. It works out to about 5.5 trillion days.

People sometimes ask, "What happened one second before the Big Bang?" This question is probably unanswerable, but it makes sense to think of time and duration as being defined relative to our universe. It is conceivable that ours is the only universe that has ever existed, or it could be one in a series of universes. If there was a universe before ours, then it would have had its own definition of time and duration, but that definition would be independent of our own. If there was no other universe before ours, then the concept of time would be undefined until our universe came into being. Either way, using our concept of time, there is no meaningful way to talk about time before the Big Bang.

Calendars to Measure the Days, Weeks, Months and Years: Capturing "Tempus Fugit"

In ancient times, people developed calendars to plan their crop and planting seasons, and to know when the seasonal weather patterns would change. As we have known since Copernicus' time around 1530, the apparent movement of the sun is caused by the movement of the earth around the sun, while the apparent movement of the moon is due to its movement around the earth. Despite the presence of many visible stars, early peoples turned to the regular movements of the moon and sun to form the basis of their calendars. The stars appeared relatively static, while the sun and moon were far larger and both had regular cycles. It takes about 28 days for the moon to revolve around the earth and about 365 ¼ days for the earth to revolve around the sun. Many different calendars were developed over the years, and some of the more significant ones are reviewed here.

The Babylonian Calendar: One early calendar was created by the Babylonians around 400 B.C. Their Metonic calendar was based on a 19-year cycle for the years. They assigned 13 months to seven of their years and 12 months to the remaining 12 years. The months were based on the cycles of the moon. While this worked out in practice, it was somewhat too complex for everyday use.

The Egyptian Calendar: Even earlier than the Babylonians, the Egyptians created a calendar that is not too far from the one we use today. Much of their life centered around the Nile River, which regularly flooded from the end of June until October. During this time, crops were planted, and grew until late February. The crops were then harvested from late February until the end of June. The Egyptians measured the flood level of the Nile with a vertical stick. This regular cycle of the Nile became the basis for a "Nile Year." The Egyptians discovered that by allocating 30 days to each month and adding five extra days at the end of the year, they had a workable calendar. The last five days were used for a festival, in which the birthdays of certain Egyptian gods were celebrated.

The Egyptians also noticed that once a year, Sirius, the brightest star visible, rose on direct line with the rising sun. This always occurred during the Nile's flooding season. As a result, they decided to use this date as the start of their year. This was the time when the five-day festival occurred. Even though the Egyptian calendar year was actually ¼ day short, it took many years for this small difference to be noticeable.

The Roman Calendar: Romulus is credited with creating the Roman calendar about 753 B.C. when he founded Rome – although a great deal of mythology surrounds his existence and the city's founding. The Roman calendar starts in March and originally had ten months. Six of these months have 30 days and four have 31 days. This made for a 304-day year calendar.

This calendar was not very practical because it was not aligned with the seasons. Around 700 B.C., it was reformed by adding the months of January and February. This increased its length to 355 days. This was still 10 days short of the solar year. As a result, an extra month was added in some years to make up the difference.

The Julian Calendar: The Julian calendar, used in most of Europe and in European settlements in the Americas until the Gregorian Reform Calendar replaced it, dates back to Julius Caesar in 46 B.C. This calendar was very similar to the Egyptian calendar, and attempted to fix some of the issues with the Roman calendar. In the Julian calendar, every year averaged out to 365 ¼ days. This was accomplished by introducing a Leap Day every four years. To make up the difference between the Julian and Roman calendars, Julius Caesar added 90 days to the Roman calendar in the year 46 B.C. between November and February.

The Muslim Calendar: The Muslim calendar, also known as the Islamic calendar and the Hijri calendar, is quite different from the Gregorian calendar. It is designed to be in tune with the phases of the moon, not with the solar year. The Muslim calendar is based on a cycle of 30 lunar years. Each cycle contains 12 lunar months that start on or near the new moon. Nineteen of the 30 years have 354 days, while the remaining 11 have 355 days. This makes the Muslim years 10 or 11 days short of the solar year. The first Muslim year began on Friday, July 16, 622 A.D. in honor of the flight of the prophet Mohammed from Mecca to Medina. The year 2014 was the year 1435 in the Muslim calendar.

Ramadan is one of the major events in the Muslim calendar. It is a month of fasting when food and drink cannot be taken during the day until sundown. Also, Muslims are expected to pray at regular times during this period. The timing of Ramadan is based on the Muslim calendar. It begins at the sighting of the crescent moon with the unaided eye on the ninth lunar month. Ramadan lasts for a lunar month. In 2014, it began on July 28, and continued to July 28. In terms of the Gregorian calendar, Ramadan tends to start about ten days earlier each year.

The Gregorian Reform Calendar: Despite the improvement of the Julian calendar over the Roman calendar, and despite the addition of leap years, the Julian calendar was still not quite in tune with the solar year. The solar year is actually 365 days, 5 hours, 48 minutes and 46 seconds long. This calendar, inaugurated in 1584, eliminates leap years in those years that are divisible by 400, such as 1600 and 2000. The Gregorian calendar is in tune with the solar year almost completely, and is off by only a fraction of a second per year.

It was the Gregorian calendar that introduced the term A.D., meaning Anno Domini, which is Latin for The Year of Our Lord. The term "A.D." is used to refer to years dated after the birth of Christ, while B.C. means "Before Christ." A date such as 2015 A.D. literally means "the 2,015th Year of Our Lord." The term "A.D." is usually assumed in designating the year, unless there is the potential for confusion with a B.C. year.

In order to make up for the incorrect results of the Julian calendar, Pope Gregory XIII did a shocking thing that could never be done today: He decreed that in 1582, October 4 would be followed by October 15. And so 10 days from the calendar forever disappeared. This did not sit well with servants, who demanded their usual full monthly pay. No doubt landlords took the same position about their rent. But the positive effect of this change, when combined with the minor change in eliminating leap years divisible by 400, resulted in a calendar that was once again in tune with the solar cycles and the seasons. This is essentially the calendar that is still used in Western countries today.

Because the Gregorian calendar came from Rome, Protestant England and the Protestant colonies in America refused to go along with this change. They finally agreed to go along with the Gregorian calendar in 1752, beginning the New Style year on January 1. After 11 days were added to the calendar, George Washington's birthday suddenly got moved from February 11, Old Style, to February 22, New Style. No doubt many similar adjustments had to be made. In the end, though, the change took effect in England and America, and it stuck.

The Evolution of Clocks and Other Time-Keeping Devices: "Let Not the Sands of Time Get in Your Lunch"

In ancient times, there was no need for the small units of time we have today. Life was largely based on agriculture, and the most important elements of time were the change of seasons and the length of daylight time when work could be done. These changes between seasons and between day and night could be observed without clocks or watches.

The earliest attempt to measure time was by means of sundials, or shadow clocks (Figure 5-1). These devices had a vertical bar that cast a shadow, the length of which varied with the position of the sun in the sky. As the sun rose in the heavens, the shadow became shorter. The sundial was reversed at noon, or the sun's high point, and the shadow became longer as the sun descended. The use of these "sun clocks" goes back at least to the ancient Egyptians, and can be dated to 1500 B.C.

Figure 5-1. A sundial from Suffolk, England

Of course, one problem with sundials was that they only marked the hours during the day, when the sun was shining. The next step in timekeeping was the use of water clocks. Even the ancient Egyptians used water clocks. The passage of time was measured by the water level in a bowl that had a single hole at the bottom from which water dripped. The water clock was marked with lines that divided the water clock into units of time that could be read in the darkness of night.

Figure 5-2. Two early Greek water clocks

The Greeks (Figure 5-2) used water clocks to set limits to the time allowed for pleading in court. Six minutes was a common length of time allowed for pleading. Apparently it was not uncommon for orators to ask that the water flow be stopped temporarily so they could have more speaking time. This sounds somewhat like the Greek equivalent of the modern filibuster in the United States Congress.

The Romans used both sundials and water clocks to keep time. The sundials were used on sunny days, while the water clocks marked the time on cloudy days and at night. Foreshadowing our contemporary watches, they used portable sundials, measuring 1½ inches across, for carrying about in a pocket. However, because of the seasonal variation in the length of the days, it was difficult for the Romans to find agreement on the exact time. Like the Greeks, the Romans also used water clocks to time their speeches. And like the Greeks, they tried to extend their speaking time with these clocks, at times requesting multiple water clocks.

The next step in the measurement of time came with the hourglass in the 8th century A.D. The hourglass measured time with flowing sand rather than dripping water. The hourglass was useful for measuring a specific amount of time, but was not suited to measuring out fractional portions of that time. It was also not very convenient for keeping track of time for an extended period of time during the day or night, since it had to be turned over repeatedly as the sands of time ran out. But hourglasses were used for measuring shorter time periods. By the 15th century, they were used in England to measure the lengths of preachers' sermons, and in the 16th century, they found their place for measuring short intervals of time in the kitchen.

The Rise of the Equal Hour and the Mechanical Clock: "I'm Late! I'm Late! for a Very Important Date!"

It is important to be aware that clocks do not sense time in the way in which our eyes sense light or our ears sense sound. Instead, whether they rely on sundials, dripping water or flowing sand, they rely on natural forces to operate in parallel to the flow of time. Through the use of markings that designate units, clocks enable us to create units of time. These units are arbitrary (although agreed upon), except that they are designed to capture the passage of day and night and the duration of each.

While sundials and water clocks served the function of timekeeping for a number of centuries, eventually additional needs arose that could not be satisfied by these devices. One such need was not so much to mark the hours as to sound the time for the purposes of the community and to announce the time for church services. In the 14th century, churches and town halls had mechanical clocks that sounded the time each hour. These clocks were designed to repeatedly stop and then release a falling weight. The falling weight was connected to the operation of the clock's machinery. With the invention of mechanical clocks, time became divorced from sun cycles and flowing water, and instead began to be measured by the movement of mechanical weights that eventually would mark off small units of time.

Clocks and the Equal Hour

Initially the clocks struck the hour, but did not necessarily have markings that divided the day into hours of equal length. But by the 1500s, some clocks had dials that marked the hours and even quarter hours. Initially these quarter hours were sometimes marked 1, 2, 3 and 4. Eventually these were replaced by 15, 30, 45 and 60 to indicate the minutes.

Christian Huygens, a Dutch astronomer and mathematician, is credited with inventing the first pendulum clock in 1627. By the mid-1600s pendulums were commonly used to regulate the movements of clocks. This made it possible to add a minute hand and eventually a second hand.

How we ended up with a 24-hour day divided by hours, minutes, and seconds seems to be a result of the design of early sundials and of geometry. Because of their design, it was natural to divide sundials into equal units or segments, and many early sundials were divided into twelve equal parts. So in some ways the segments on a sundial resemble today's clocks, with 12 equal parts dividing the day into 12 hours. However, sundials were not effective at night, so the second 12-hour period could not be marked with the sundial. Once mechanical clocks were invented, it became possible to create clocks that divided the day into 12 equal units. Today the 12-hour periods are divided into a.m. and p.m., making for 24 hours total. Our term "hour" originally meant one-twelfth part of the sunlight or darkness.

Our use of 60 minutes and 60 seconds seems to be rooted in Babylonian geometry. The Babylonians based their number system on the number 60, and marked off 360 degrees in a circle. The Egyptians used 360 as the number of days in the year, plus five more at the end. They also divided a circle into 360 degrees. The number 60 was 1/6 of 360, and became a convenient subdivision of each hour. Once the hour was divided into 60 segments, or minutes, it was natural to use the number 60 as a subdivision of the minute. The term "minute" is derived from Latin and means "small division." It still retains this meaning today, meaning "small amount," though the accent is on the second syllable. The term "second" means "second minute."

It is a fascinating fact of history that so often progress consists in taking something that is manual or mechanical and improving it through automation without changing its fundamental nature. For example, today's computers are modeled on the old manual typewriters. Typewriters went from manual to electric to Selectric (IBM's electric typewriter) with memory. Eventually early computers using punched cards were invented. Then the memory got separated from the keyboard entirely, and electronic computers were invented. First they were very large, then desktop models were invented, followed by the personal computer. Then personal computers became portable as laptops, and then tablets. All these changes were made without fundamentally altering the keyboard in some form as input. Only tablets began introducing input apart from the keyboard.

During the entire time the computer was evolving, no one thought to look at the most fundamental aspect of the computer in terms of entering data: the keyboard. The typewriter keyboard with its anagram of letters designed by a left-handed person became so established that it could not be changed. For example, the most commonly used letter, 'e', is easily reachable by the left hand. This keyboard remained intact, only now it became electric or electronic. This remains true today, although some speech recognition programs exist that may finally make it possible to bypass the keyboard.

Much in the way today's computers with their keyboards automated the mechanical typewriter, so our system of electronic time today is rooted in sundials and ancient geometry. Although the origins of our 24-hour day remain somewhat obscure, it most likely developed from putting two 12-hour days together. And these 12-hour days could be traced to the divisions or markings on a sundial, since the average length of a day was 12 hours during a good part of the year. The division of the hour into 60 minutes goes back to the Babylonian geometry that marked a circle into 360 degrees. As noted above, 60 is 1/6 of 360, and with 60 minutes making up an hour, it followed naturally to have the "second minute" equal 60 seconds. Even though our clocks and watches today are electric or battery operated, they still demarcate the same units of time first marked by primitive sundials.

Clocks Continue Their Development

After the invention of the pendulum clock by Huygens in 1627, clockmakers focused on developing more accurate and useful clocks over the rest of the century by making modifications to the pendulum and to the clock case. In 1675, Huygens invented the spiral balance spring. This spring controlled the rotary oscillation of a balance wheel inside a clock, and proved especially useful for watches. The spiral balance wheel performed the same function as a pendulum, in that it controlled the clock's movements, but did so in a much more compact way. During the next century, clocks became more accurate and watches became more popular.

In the early 1800s, the attention of clockmakers turned towards making them less expensive and towards methods of mass production. While European watchmakers dominated the market in the 1840s, an American company helped get clock making established in the United States at the same time. The Boston Watch Company was established in Waltham, Massachusetts during this period. In 1853, the company produced 1,000 watches, and by the end of 1854, it was producing 36 watches per week. The company later produced watches for the Union Army during the Civil War.

High precision clocks became the focus of clockmakers at the beginning of the 1900s. A German engineer developed a regulator housed in a partial vacuum that increased the accuracy of pendulum clocks. In 1928, an engineer at Bell Laboratories discovered the quartz crystal as a means to achieve high accuracy in clocks. A quartz crystal vibrates at a highly regular rate when an electric current is applied to it. By the end of World War II, quartz crystal clocks had improved to an accuracy variation of just one second in 30 years.

Despite the accuracy of quartz crystal clocks, they were displaced as a time standard beginning in 1948 with the atomic clock. Atomic clocks based their timekeeping ability on an atom's resonant frequency. This refers to the periodic oscillation of the atom between two of its energy states. Further work in the 1950s led to the cesium beam atomic clock. This led to the adoption of a new standard of time in 1967: a definition of the second based on the resonant frequency of the cesium atom. One second is defined as the duration of 9,192,631,770 periods of the radiation corresponding to the transition between the two hyperfine levels of the ground state of the cesium 133 atom.

A Change in the Conception of Time: "He's So Slow that He Takes an Hour and a Half to Watch '60 Minutes'"

It is difficult to define time, but time is really a measure of duration. The universe endured billions of years before time was measured. In relation to the age of the universe, human beings have been here for only a very brief time. However, the way in which humans measure time influences how it is conceived of. When time was measured with flowing water and sand, for example, it was natural to conceive of it as a flowing medium. Once mechanical clocks came along to measure time, it became natural to think of it as a set of discrete moments. Thus the advent of the mechanical clock forever changed our conception of time.

Decimal Time

Looking back at the origins of our current time system, as we have seen it dates back to the Babylonians and Egyptians. The Babylonians used mathematics with a base of 60, and divided a circle into 360 degrees. The Egyptians divided hours into 60 minutes and minutes into 60 seconds. They also began measuring time with sundials and water clocks.

Decimal time offers a different picture. While traditional time divides the day into 24 hours, many decimal time systems divide the day into 10 hours. At the same time, decimal time allocates 100 minutes to each hour and 100 seconds to each minute. The two time systems are compared in Table 5-1.

Table 5-1. A Comparison of Traditional Time and Decimal Time

Traditional Time	Decimal Time
24 hours per day	10 hours per day
60 minutes per hour	100 minutes per hour
60 seconds per minute	100 seconds per minute
1440 minutes per day	1000 minutes per day
3600 seconds per hour	10000 seconds per hour

Decimal time has never actually been in effect for a broad population, with the exception of the experiment tried by the French just after the French Revolution. On November 24, 1793, the French instituted decimal time within France. French Revolutionary Time was based on a 10-hour day, with 100-minute hours and 100-second minutes. They manufactured clocks like the one shown in Figure 5-3 to ease the transition.

Unfortunately, people found it very difficult to make the adjustment to decimal time. In addition, replacing every clock and watch in the country proved to be too expensive and too much of a change to keep the new system going, and it lasted only 17 months.

Figure 5-3. A French Revolutionary Time Clock showing both traditional time and decimal time in one clock.

Decimal time does have some advantages over traditional time. Forty percent of a 24-hour day is 9.6 hours, or 9 hours and 36 minutes. In decimal time, 40 percent of one 10-hour day is 4 hours, or 400 minutes.

So much of our mathematics is figured on a base ten system that converting fractions to the 24-hour 60 minute-system can be awkward. For example, it is fairly standard in filling out timesheets for payroll to report time in decimal units (e.g., 21.5 hours or 18.75 hours). But 21.5 hours has to be converted to 21 hours and 30 minutes, while 18.75 hours is really 18 hours and 45 minutes for payroll purposes. In a decimal system, 21.5 hours would be 21 hours and 50 minutes, while 18.75 hours would be 18 hours and 75 minutes.

It is interesting that much of the world has converted to the metric system for weights, volume, and measures, but has resisted the transition to decimal time. In many places except for the United States, pints and gallons have been replaced by liters and the meter has replaced the yard as a unit of measure. Likewise, kilometers have replaced miles as a measure of distance in most locations, especially in Europe. Even the United States is in the process of converting over to the metric system for many measurements. Yet there appears to be no move toward decimal time, despite certain obvious advantages. Some of these advantages are described in the next two sections.

Flowtime: An Alternate System Based on Decimal Time Since "Time Waits for No Man"

Flowtime is a system of decimal time that contains elements of the French decimal time but retains some elements of traditional time. A transition to flowtime would therefore be much easier than the transition the French tried to impose in the late 1700s. In another sense, flow time today is defined as "the period required for completing a specific job or a defined amount of work." This is a different use of the term "flow time" that is only tangentially related to the use of the term "flowtime" in this chapter.

It is surprising that after 3,300 years, we are still operating on a system of time that was invented long before technology and 2,600 years before the invention of mechanical clocks (around 1300). Today we have many reasons to divide time into smaller and smaller units. Flowtime recognizes this, and it offers a system of time that harmonizes much better with our numbering systems in other areas of life. Most of these are based on the idea of ten. Decimal systems are intuitive because people find counting to ten on their fingers to be intuitive.

The proposal for flowtime is to switch the counting of minutes and seconds from 60 divisions to 100 divisions. This proposal does not include any change in the number of hours per day. It only proposes to increase the number of minutes in one hour from 60 to 100. Likewise, it increases the number of seconds in a minute from 60 to 100.

To easily convert from traditional time to flowtime, take the minutes or seconds in traditional time and multiply by 5/3 or 1.67. The result is the minutes or seconds in flowtime. The hour remains the same.

Another easy way to make the conversion is as follows: Take the minutes or seconds in traditional time and multiply that figure by 2/3. Then add that value to the traditional time value, and you have the flowtime value. For example, if it's 1:15, take 2/3 of 15, which is 10. Add 10 to 15, and you have the flowtime of 1:25.

What are the implications of this? It means that, under flowtime, instead of the time being 1:30 pm, it will be 1:50 pm. Instead of 3:45 pm, the time will be 3:75 pm.

Why Change to Flowtime?

There are several good reasons for changing to flowtime:

1. **Flowtime divides time up into smaller quantities.** Instead of 60 minutes per hour, there are now 100 minutes. Instead of 1440 minutes per day, there are now 2400 minutes per day. Instead of 3600 seconds in one hour, there are now 10,000 seconds per hour. Having smaller units of time to deal with makes it possible to break tasks down into more discrete periods.
2. **The advent of digital time makes the base 60 method of measuring time obsolete.** When the only types of clocks were analog clocks, base-60 type clocks made more sense. With the advent of digital clocks, counting down from one minute 20 seconds to 59 seconds introduces a gap as the time reaches the one-minute mark. It would be more intuitive to go from 101 to 100 to 99 seconds than to go from 1 minute 1 second to 1 minute 0 seconds to 59 seconds.
3. **Flowtime provides a more fine-grained analysis of time for sporting events.** A basketball or football game played on flowtime would have that many more time units built into it. While it will not actually make the game last longer, the possibility for additional plays is increased because the unit of time is smaller. In fact, at National Basketball Association (NBA) games, as the clock winds down, the seconds get divided into ten equal segments. As a result, if a foul occurs with 0.3 seconds left on the clock, it is possible to stop play at that point. If the second were not subdivided into ten equal parts, no one would know exactly how much time was left. With this form of decimal time in place, players know they may still have time for a "catch and shoot" play after they get the ball. The same idea applies in daily life.
4. **The advent of computers and other time-oriented equipment makes it**

necessary to measure time in ever smaller chunks. Computer time is now measured in nanoseconds. While we don't need to measure our ordinary time in nanoseconds, flowtime gives the option of having a more fine-grained analysis of time.

5. **Many time accounting systems are based on decimal time.** As has already been noted, timesheets are often made out in a form of decimal time.
6. **Here's an analogy that will help explain the value of flowtime.** Some coffeepots have markings for 2, 4, 6, 8 and 10 cups. For someone who wishes to make 5 cups, it would be helpful also to have markings for 3, 5, 7, and 9 cups, so it isn't necessary to estimate what is halfway between 4 and 6. Flowtime is like a coffeepot with extra markings — it enables you to measure time to a higher degree of precision.
7. **Also imagine measuring with a ruler that only has the 1/4 inch and 1/2 inch markings on it.** If you want to measure something that is 4 3/8 inches, you will have to estimate the halfway point between 4 1/4 and 4 1/2. If you then switch to a ruler that has the 1/8 and 1/16 points marked off, you can make a more precise measurement. Flowtime is like a ruler of time that gives you more precision than our current time system.

Where's the payoff in this switch to flowtime? Why does it matter what time system we use as long as everyone has the same one? **The payoff in switching to flowtime is that when you switch to flowtime, you will have the tools for becoming more productive.** The reason is quite simple. Because you are working with smaller time units, you have the potential to achieve greater precision in the measurement of time.

Someone who gives himself one hour to complete a series of tasks, such as writing five letters, is less likely to get them done than someone who gives himself 12 minutes to complete each letter, and monitors how long each letter takes. The first person is likely to find himself rushing at the end to complete the five letters, unless he paces himself along the way. Under flowtime, the same person can allocate 20 minutes to each letter. Of course, the duration of 12 minutes of traditional time is the same as the duration of 20 minutes of flowtime, so flowtime may not be that much of an advantage in this case.

On the other hand, consider the example of someone that has 25 things to do in an hour, such as getting 25 letters ready to mail. To accomplish this, each letter has to take about the same amount of time, including signing the letter, writing a note, putting in the inserts and sealing the envelope. Under traditional time, our letter writer has 2.4 minutes for each letter. However, under flowtime, he has 4 minutes per letter. If he has access to a flowtime clock, he can easily track his progress by the flowtime clock. With a traditional clock, he would have to approximate the time each letter takes. In this case, flowtime provides an advantage by dividing time into smaller units, even though the duration of time remains the same.

The advantage of flowtime was driven home to me again one evening while watching a National Basketball Association (NBA) game. There were 3.5 seconds left. The commentator said, "The coach wants to know exactly how much time is left so he knows what plays he can run." When the game is on the line, the coach has a choice of plays he can run to try to score a basket in a very limited period of time. One might take 3.1 seconds, while another might take 3.9 seconds. Because he knows that there are exactly 3.5 seconds left, he knows to call a play that can be completed in 3.5 seconds or less. Without the decimal time clock, he would only know that there are somewhere between 3 and 4 seconds left in the game, but he wouldn't know exactly how much time is left.

By dividing time into smaller units, flowtime enables us to determine more exactly how much time is available, when this is desirable. And the change can be made without abandoning the 24-hour day, unlike French Revolutionary Time, which called for a 10- hour day.

Figure 5-4 shows the faceplate of an analog clock that depicts both conventional time and flowtime. In this clock, the short hour hand would point to the hours on the inside circle, while the long minute hand would point to the minutes on the outside of the circle. So at 2:30 in conventional time, the short hand is exactly halfway between 2 and 3, while the long (minute) hand points straight down to 50. In conventional time it is 2:30, while in flowtme it is 2:50.

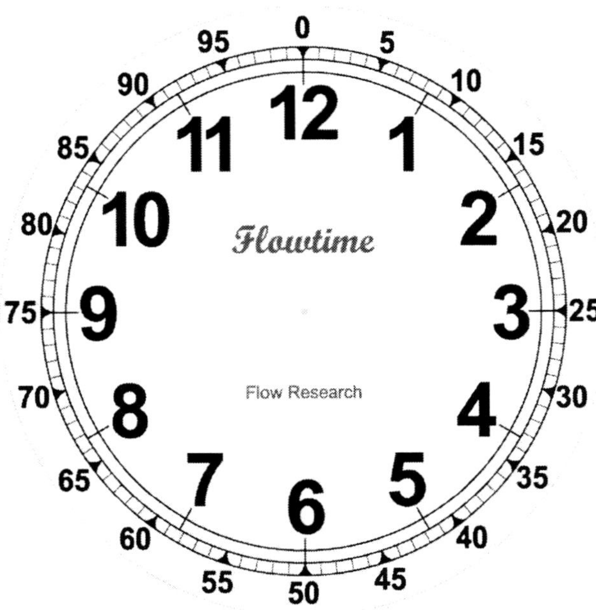

Figure 5-4. The face of a clock showing both conventional time and flowtime

Table 5-2 gives approximate equivalencies between traditional time and flowtime for the 2:00 hour. Of course, the same conversions can be used for any hour.

Table 5-2. Conversion of Traditional Time to Flowtime Minutes

Traditional Time	Flowtime	Traditional Time	Flowtime
2:00	2:00	2:31	2:52
2:01	2:02	2:32	2:53
2:02	2:03	2:33	2:55
2:03	2:05	2:34	2:57
2:04	2:07	2:35	2:58
2:05	2:08	2:36	2:60
2:06	2:10	2:37	2:62
2:07	2:12	2:38	2:63
2:08	2:13	2:39	2:65
2:09	2:15	2:40	2:67
2:10	2:17	2:41	2:68
2:11	2:18	2:42	2:70
2:12	2:20	2:43	2:72
2:13	2:22	2:44	2:73
2:14	2:23	2:45	2:75
2:15	2:25	2:46	2:77
2:16	2:27	2:47	2:78
2:17	2:28	2:48	2:80

2:18	2:30	2:49	2:82
2:19	2:32	2:50	2:83
2:20	2:33	2:51	2:85
2:21	2:35	2:52	2:87
2:22	2:37	2:53	2:88
2:23	2:38	2:54	2:90
2:24	2:40	2:55	2:92
2:25	2:42	2:56	2:94
2:26	2:43	2:57	2:95
2:27	2:45	2:58	2:97
2:28	2:47	2:59	2:98
2:29	2:48	3:00	3:00
2:30	2:50	3:01	3:02

Units of (Conventional) Time

1 nansecond = 1/1,000,000,000th of a second

1 microsecond = 1/1,000,000th of a second

1 millisecond = 1/1000th of a second

1 minute = 60 seconds

60 minutes = 1 hour

24 hours = 1 day

7 days = 1 week

1 week = 168 hours

1 week = 10,080 minutes

1 week = 604,800 seconds

1 year = 365.25 days

1 year = 8,766 hours

1 year = 525,960 minutes

1 year = 31,557,600 seconds

Morley's Point:
Time

In this chapter, *Measuring Time As It Flows On*, Jesse is trying to get us "unstuck" from the mud of several thousand years of clumsy analysis. You will find that sections, such as volume, length, and others offer a degenerative coupling to the past. Physicists used to say "shut up and calculate" was the best way to go – not so anymore.

I recently purchased a clock to fight back against my finance people. The clock has some algebra and arithmetic on it and should confuse the beanies. It uses pi^2 to define nine o'clock, giving the math guys a migraine. We are using techniques the way some guy told us to do it in 400 B.C. Jesse is trying to change all that. I was impressed indeed by his ability to measure circles and not use sticks.

Jesse also does a good job of explaining the history of time and how it made a new dimension of perception for us humans. Time, when you are an animal, is measured in terms like "I'm warm," "I'm cold," or "I can't see," but even animals have the perception of sequence in sensory measurements. My dogs wake me up at 7:20 AM every day. The conception of time itself, whether seasons or nanoseconds, is the root real invention here. We need to use Jesse's concepts to make sense out of the time we're now using and may be using later.

I do have some objections to the term "new time." Society will not bend that easily. Engineers and logical thinking usually make the mistake of excluding the social aspects of change.

In the past, time was measured using natural events. These events included such things as sunrise and sunset, planting time (winter and summer) and how the moon changed with observation. The saying "many moons ago" is a time measurement using natural events. However, as human knowledge advanced, natural events were not sufficient to narrow down time measurement, so we had to make artificial events. The ticking of a clock is one of those events. A squirming cesium atom is another such event, as is the hour-long class in academia. We made events to get a higher resolution of observational time.

This chapter well describes time near the "moving now." Jesse helped me here with my understanding of time. The future of time is something else again. Your GPS system needs accurate quantum corrected time to find the position of your car with respect to the environment. The ability to make accurate, consistent events is what we call time. Our observation of clock ticks (or a pendulum swing) gives us a reasonable time for most human situations. For example, a natural ticking clock of an automated system is the power being fed to it. It comes every 8 to 10 millisecond in lumps of energy – 60 cycles per second. Time rationale need not be faster and more accurate than that for measurement of the system. There are components in the system that require more accuracy, but for engineers, a slide rule is just as good as a calculator.

There are some wildcards, however. Most are outside the scope of this book, but time itself is variable. A photon travels from the sun to the earth in zero time as measured by the photon brain. Time is also relative. I recently read the book titled *My Brief History* by Stephen Hawking. He started talking about imaginary time – egad. I started to think about that. In the power analysis world, we can use imaginary numbers to analyze power as a vector. For example, if we have current without voltage, no power is delivered, or if we have just voltage without current, again, no power. The power line itself can have a voltage/current relationship that is nonsynchronous. In a machine shop, capacitors are used to optimize power use, and so it is with time.

Real-time can be described as events per unit of time. The vertical axis can be the observation count and the event count, with the horizontal axis being time. Many events occurring that are not observed means that, for the observer, no time has passed. If many observations have been made, but no events are occurring, for the observer no time has passed. So now you can vectorize our time analysis on a two-dimensional basis. I don't believe it either, but it's interesting to talk about.

The mystics take another road. For instance, something is timely; make time for supper. The velocity of time in the mystical sense is beyond my struggling left brain's ability. Einstein said, "The distinction between the past, present, and future is only a stubbornly persistent illusion." I'm surprisingly pleased with the headaches given by this discussion. On a personal note, this chapter was the most difficult for me. I couldn't weigh, measure or bottle this "time." It's all imaginary anyway.

"It is of great use to the sailor to know the length of his line, though he cannot with it fathom all the depths of the ocean." – John Locke

Chapter Six

Going to Great Lengths in Measurement

The concept of length is one of the most familiar in our experience. Being able to accurately measure length is necessary for many of our most common tasks. Imagine building a house without being able to measure and specify the lengths of the cinder blocks and pieces of wood used to construct the foundation and structure of the house. Building furniture is another area where exact length measurement is necessary for good results. Today's sophisticated industrial environment also requires precise measurement of the length of parts for cars, machines, pipes, and even computers.

John Locke

While there are many possible units for measuring length, most people seem to be able to adjust their use of the various units to the degree of precision desired. For example, no one is likely to complain about a road sign to a city reading "Boston – 10 miles" if it isn't exactly 52,800 feet to Boston, Massachusetts from the sign. These road signs are meant to give the distance to the center of the city, but even that may not be a precisely defined area. And if it is actually 9½ miles to the center of Boston from the sign, no one is likely to complain.

In other cases, precision is a lot more important. Someone who orders a desk that is supposed to be 28 inches high and gets one that is 30 inches high can justifiably complain, especially if the desk is intended to be used for a computer. And two sides of a desk that aren't quite the same length can result in a desk that won't have a flat top, and may not stand straight. This is just one of the many examples in ordinary life where precision in length measurement is important.

Defining Length: Can You Please Hold the Other End of This Rule?

Length can be defined as the distance between two points. Saying that the length of room is 8 feet is equivalent to saying that the distance between two points at either end of the room is 8 feet. If someone holds a rule to measure the length of a room, the endpoint of the measurement is the mark on the rule that delineates the number of feet and inches that coincides with the edge of the room. And the beginning of the measurement is the beginning point of the rule.

Uniting on a Unit

What is essential to measuring length is having a unit of measurement. The American system uses the units of inch, foot, yard, and mile. This is also called the imperial or U.S. system of measurement. The units are derived from English (British) units. This system is still in use in parts of the United Kingdom (UK), although the metric system is used in the UK for much of the measurement in scientific and industrial contexts. Likewise, the metric system has made inroads in the United States, but the old imperial units still dominate.

The metric system uses the centimeter, meter, and kilometer as the fundamental units of length, and is the one most commonly used in countries outside the U.S. The metric system originated in France in the 17th and 18th centuries. It is now called the System International of Units and is abbreviated SI. It is based on multiplying or dividing its basic units by ten and its powers.

The Importance of Establishing a Standard of Measurement

While the units of length vary from one society to another, they are derived in a similar way. In order to establish a unit of measurement, whether it is an inch, a foot, a meter or any other measurement unit, the unit has to be defined in terms of some known standard. Unless the unit is defined in terms of a standard, it will not have a fixed reference and cannot be used as a unit of measurement. The standards for the units of measurement we use today have changed many times over the centuries.

U.S. Standard and the Evolution of English Standards

In ancient times, units of measurement were often defined in terms of parts of the body. This was convenient in that it provided a standard readily available to all. The obvious drawback was that these units varied from one person to another.

It is generally accepted that the standard for a foot was the length of the average man's foot, although the 12-inch foot as defined today is somewhat longer than the length of the modern human foot. The inch was commonly defined in terms of the width of a man's thumb. The term "yard" was derived from the Old English word "gyrd," meaning "rod" or "measure." The yard came to be known as the distance from the nose to the end of the middle finger of the out-stretched hand. In the 12th century, King Henry I of England decreed the yard to be the distance from his nose to the tip of his out-stretched thumb. This closely approximates the distance we call a yard today.

The definition of a foot as consisting of 12 inches goes back to Roman times. When the Normans conquered England in 1066, they brought with them the Roman tradition of measurement of using a foot with 12 inches. At some point there was an attempt to standardize the length of a foot. Some attribute this to Henry I, who reigned from 1100 to 1135. Later in the 1100s, a "foot of St. Paul's," which was similar to our modern foot, was inscribed at St. Paul's church in London at the base of a column so all could see. During this time the idea arose of defining the length of a foot as being 1/3 yard, making the yard the primary unit of measurement. This idea appears to have come from Henry I, who is said to have ordered the construction of three-foot standards, called yards.

With the yard as a primary unit of measurement, standardizing its length became more important. In 1588, the standard yard of Elizabeth I was constructed. It consisted of an iron bar with a square cross-section, measuring about ½ inch on a side. In the 1700s, the emergence of industrialization made it more important to have standardized lengths. In 1742, the Royal Society in London arranged to make an exchange of standards with the French Academy of Sciences. The Royal Society had two identical brass bars made, based on the length of a yard measured in 1720 according to the standard yard of Elizabeth I. The two bars were sent to the French, who marked off the length of half a toise on the bars and sent one back. A toise was at that time a French unit with a length of six feet. This is one of the earliest attempts to standardize the definition of a yard with respect to the length of a metal bar. It built on the earlier attempt made in 1588.

Later in the 1740s, the British government decided it was time to establish a more exact standard yard. A 63-member commission was appointed to decide how to proceed. An instrument maker named Mr. Bird was appointed to make a more exact version of the standard bar that was exchanged with the French in 1742. Mr. Bird made two such bars, one in 1758 and one in 1760. Neither of these standards was officially adopted until May 1825, when a legislative act declared the bar made in 1760 to be the official yard standard. This yard remained the English standard until May 1834, when a fire burned down the Houses of Parliament and destroyed both of Bird's standard yards. Though at the time there was no other means available, this shows the danger of relying on a perishable physical object to define a measurement standard.

After the destruction of the standard yard, a new committee was formed in 1838 to create a new standard. By this time, levels of precision in measurement had advanced considerably. The committee appointed Francis Bailly to determine the best metal alloy to use. He designed the form of a new standard bar before he died in 1844.

The task of creating the new standard was assigned to a Reverend Sheepshanks. By this time, length could be measured to the precision of one part in ten million. New thermometers were created that were accurate to the 1/100th of a degree to account for thermal expansion. The standards were measured floating in a pool of mercury to eliminate the effects of bending. Over a period of years, he created 40 bars in an attempt to find one that most closely matched the measurement of the second of the two bars that were destroyed in the fire. In doing so, he used the guidelines established by the standard yard committee formed in 1838. Eventually one of these was selected, and it was declared the legal standard in 1855.

In the United States, parallel developments were occurring, although along slightly different lines. The Treasury Department recognized the yard as the basic unit of length for customs purposes in 1832. The standard yard in the United States was a brass bar sent to the United States in 1815 by an instrument maker in London.

In 1856, the British government gave the United States two of the 40 new standard bars that had been constructed following the destruction of the earlier standard bar in the fire. One of these bars was bronze and one was iron. The bronze bar was selected as the new standard yard, although it was 0.00087 inches longer than the brass bar from the London instrument maker that was declared the legal standard in Britain in 1855.

Today's Definition of Meter

Since 1855, the science of metrology has made major advances, with a greatly expanded capability for precision. In 1866, an Act of Congress made the use of the metric system permissible in the United States. In 1889, the United States received a prototype Meter bar from the International Bureau of Weights and Measures, and that became the new standard for the length of a meter. However, in 1960 the science of measurement took a major step forward with the definition of the meter in terms of the wavelength of light from krypton-86. A meter was defined as 1,650,763.73 wavelengths of the orange-red radiation of krypton-86. This definition was promulgated by the French organization called General Conference on Weights and Measures (CGPM).

In 1959, just prior to this definition of a meter, Australia, Canada, New Zealand, South Africa, the United Kingdom and the United States agreed on the length of an international yard, defined in terms of the meter. The international yard is equal to 0.9144 of a meter. Thus, the length of the yard came to be derived from the definition of a meter.

In 1983, the CGPM took another step towards precision by defining the meter as equal to the path length traveled by light in a vacuum during a time interval of 1/299,792,458 of a second. History indicates, however, that this is probably not the final word on the subject.

Even Precision Has Its Limits

When people make a measurement, they are implicitly relying on a unit of length measurement, whether it is inches, miles, meters or kilometers. And that measurement unit has to be defined in terms of a standard for that unit, otherwise the measurement unit cannot serve as an objective standard of measurement. When the human foot was the standard by which the length of a foot was judged, but there was no objective foot length to serve as the standard, the foot unit referred to different lengths. When a "standard" foot length was inscribed at St. Paul's Church in the 1100s, this was a first step toward defining an objective standard for the length of a foot.

The concepts of length and distance are closely related. Length is the measurement of the distance between two points. As long as the measurement is being made in feet or inches, then feet or inches are the smallest allowable units of measurement for the purpose of this measurement.

Of course, once the length of a foot was defined in terms of the length of a yard, this provided a new standard of measurement for the length of a foot. But we have already seen that there were some discrepancies in the length of the standard yards, due to differences in manufacturing processes and due to effects like temperature and pressure on the metal rods that served as standards for the length of the yard.

Every measurement is made to some degree of precision. The term "precision" refers to the number of digits to which a measurement is made. A measurement is not absolute; it is relative to the unit of measurement and to the degree of precision required for the measurement. While there is no theoretical limit to the degree of precision of any measurement, this does not mean that the degree of precision attainable is infinite. It is simply unlimited in the sense that it can be taken as far as the need requires and technology allows. It is not infinite in the sense that the precision of measurement can go to infinity. This distinction becomes important when considering Zeno's Paradox and certain paradoxes involving the number line. These paradoxes are discussed below.

Discrete vs. Continuous

The difference between the discrete and the continuous is so fundamental that it is difficult to define in simpler terms. The term "discrete" refers to something that is a single entity or object. For example, a cup, a car, a shirt, a tree and a cellphone are all discrete objects. A person can also be considered a discrete entity.

The term "continuous" refers to something that is not composed of discrete parts. What is continuous is often, though not always, in motion. For example, the flowing water in a river is continuous, and the motion of a runner going from the starting line to the finish line without stopping is also continuous. The motion of fish swimming in a river is continuous, provided they do not stop to rest. The motion of a baseball after it is hit is continuous until it stops. Time is continuous and cannot be stopped, although in certain sports such as football and basketball, a "Time Out" can be called that temporarily stops the continuous motion of the clock.

A straight or curved line is continuous, and is sometimes referred to as a continuum.

Defining the Continuum

The distance between two points is often represented as a line. This line can be straight or curved, depending on the location of the points. Many attempts have been made to define the continuum that is a line. The traditional Euclidean view is that a continuous line is made up of discrete points, and that another point can be placed between any two points. This means that a line is dense, and there is no limit to the number of points that can be drawn in the line. In fact, in the Euclidean view, a line is made up of infinitely many points. These points are conceived of as having location only, with no area.

The relation between discrete points and continuous lines typifies some mathematical puzzles that have been discussed for thousands of years and are still discussed today. Continuous lines and discrete points are fundamentally different in nature and it is not clear whether it can be correct to analyze continuous lines as being made up of discrete points, even if there are very many of them. Here the idea of infinity is brought in to be a kind of metaphysical glue. The idea is that if there are infinitely many discrete points, then these points, when packed densely enough, form a continuum, which then somehow adds up to continuity. But it is not clear how arealess points can somehow make up a continuous line, even if there are infinitely many of them.

A similar problem occurs with analyzing the area under a curve. This problem is addressed by calculus, which was independently invented by Newton and Leibnitz in the late 1600s. According to calculus, this area is determined by drawing a series of rectangles that more and more closely approximate the curved area. As these rectangles become smaller, the sum of their areas more closely approximates the curved area. As the widths of the rectangles, which become more and more numerous, approaches zero, the sum of the areas of the rectangles equals the area under the curve. This sum is conceived of as the limit of an infinite series of rectangles that are ever diminishing in width.

Zeno's Paradox

Analyzing a continuous line as being made up of infinitely many discrete points isn't necessarily wrong, and this idea has been accepted for thousands of years. It does raise some paradoxes, however, which need to be dealt with in some way. Some of these are called "paradoxes of motion," and they were formulated by a Greek philosopher named Zeno of Elea around 450 B.C. It is believed that Zeno preceded Socrates by about 20 years.

Zeno may have formulated as many as 40 paradoxes, but only 10 are known. One involves Achilles and a tortoise, and it gives an analysis of motion that purports to show that Achilles, a fast runner, can never catch up to the tortoise, provided the tortoise is given a head start.

Another paradox, well enough known to be called Zeno's Paradox, has a similar logic and involves the idea that motion is impossible. To understand this paradox, consider someone attempting to move from Point A to Point B. First, he must go halfway from A to B, which we can call A′. To move from A′ to B, he then has to go halfway again to B, to a point we can call A″. However, he is still not at B, since he must go halfway again, ad infinitum. So no one ever actually gets to point B, and hence motion is impossible.

Consider the concept of a point as a dimensionless mathematical object with no area, which is based on the Euclidean view of the number line. As discussed previously, in Euclidean geometry, a line is made up of infinitely many arealess points. It is always possible to put another point between any two points.

What makes Zeno's Paradox seem plausible is that the person who is going from Point A to B is conceived of as being located at an arealess point. Since it is always possible to put another point between any two points, he never reaches destination B.

The Prevailing Contemporary Solution to Zeno's Paradox

What is the correct solution to Zeno's Paradox? The prevailing contemporary solution relies on concepts from set theory developed by Bernard Balzono, Richard Dedekind, Georg Cantor and others. According to this view, Zeno's series of points is no longer considered to be potentially infinite. Instead, it accepts the idea of an infinite set of points and analyzes Zeno's series of points as an infinite set of points that converge to a limit. Instead of viewing the set of path lengths or points that is 1/2 + 1/4 + 1/8 + 1/16 + etc. as an infinity that can never be traversed, this analysis views it as a series that converges to 1. This idea of a set of points that converge to a limit is similar to the idea of a series of rectangles of diminishing width that converge to a limit of zero and is used in contemporary calculus to analyze the area under a curve.

An Alternate Concept: What's Your Point?

The concept of a point is fundamental to the measurement of length and distance and to Zeno's Paradox. Distance is typically conceived of as being measured between two points. For example, the nominal distance between Paris and Amsterdam by car is 510 kilometers (316 miles). This distance is arrived at by thinking of the city of Paris, France as one point and the city of Amsterdam, The Netherlands as another point.

Someone who starts driving anywhere within the city of Paris, drives 510 kilometers, and arrives at the city line of Amsterdam has measured out 510 driving kilometers between Paris and Amsterdam. However, the actual distance will vary depending on the starting and ending point. Starting at the Eiffel Tower in Paris and driving to the Van Gogh Museum in Amsterdam will produce a different total distance than starting at the Louvre and driving to the Rijkssmuseum.

What about those mileage signs that are so familiar along the highway? They presumably measure the distance from the sign itself to the city. But where in the city? Is the endpoint the city line, the city center, or somewhere else? The convention for this measurement is that the sign measures the distance from this location (the sign) to the center of the city. So a sign near Brussels, Belgium that says "Paris 209 kilometers (130 miles)" means it is 209 kilometers from the sign to the center of Paris.

What does this example tell us about points and distance measurement? Measuring distance in this way suggests that the starting point and ending point are not infinitely small mathematical points with no area. Instead, a starting point has a dimension that is relative to the type of measurement being made. If the measurement is being made with a car, the starting point is an area the size of the car that is defined in relation to surrounding objects. For example, "In front of the Eiffel Tower" is a legitimate way to define one starting point for the Eiffel Tower. There are a number of different physical positions that could satisfy this description, but they all have to be considered "in front of" the Eiffel Tower. In this example, then, a starting point has area, and its area is the size of the car that is the measuring tool. A similar definition may be applied to an ending point.

Where does Zeno's paradox go wrong? This paradox only works because the person involved is considered to be at a dimensionless point. Once the person is considered to be at a mathematically dimensionless point and at a position not taking up area, it is not possible to refute Zeno's Paradox. This is because it will always be possible to specify a dimensionless point between the point where our traveler is located and his endpoint.

Infinity, a Kind of Metaphysical Glue

Euclidean geometry neglects a simple mathematical distinction in analyzing what a line is. According to this geometry, points do not lie on the line, they lie in the line. A line, as we've said, is conceived of as being made up of infinitely many dimensionless points. Conceiving of points as lying in a line requires the introduction of the concept of infinity, since there appear to be infinitely many points in a line. Set theorists have also created different infinities of different sizes. The concept of infinity is a kind of metaphysical glue that is required to analyze a continuous phenomenon such as a line in terms of discrete arealess points. There is no way to build up to the concept of continuity when it is being analyzed in terms of discrete points without introducing the concept of infinity.

Stopping the Infinite Regress: A Practical Solution

In the example of the traveler going from the Eiffel Tower to the Van Gogh Museum, we can put an end to the infinite regress created by Zeno's Paradox because at some point, moving halfway again will not count as motion because the distance in question is too small for this prospective change. This is so because we are not defining the traveler's position in terms of position at an arealess point. Instead, the traveler's position is defined as being at the location of his physical body or vehicle. This is a three-dimensional location specified in terms of length, width and height. His actual location will usually be defined by his relationship to other physical objects around him (e.g., in the car, standing in front of the car, next to a signpost, etc.) At some point, the statement "To get to the Van Gogh Museum he must first go halfway" will no longer be true because the distance in question is so small (e.g., 1/128 of an inch) that it will be physically impossible for the car to advance such a small distance.

Once the point is reached where the car cannot move a specified small distance, Zeno's Paradox comes to an end. It is physically impossible for a human being to move his or her body 1/10,000 of an inch under ordinary circumstances. Zeno's Paradox fails with the phrase "And then the traveler advances halfway again." So one way to block Zeno's Paradox is to say that while the traveler or the car can move halfway and then halfway again, eventually it becomes physically impossible to move such a small distance.

Stopping the Infinite Regress: A Theoretical Solution

Zeno's argument has another flaw that in many ways is more serious than the one just outlined. This has to do with what it means to be at a point. Zeno would have us believe that in moving from Point A to Point B, we first go halfway, then halfway again, etc. So he breaks what is essentially a continuous line into a series of smaller segments, claiming that we never reach the end of the line. This is like someone on a trip to Amsterdam stopping at a rest stop outside of Brussels, then stopping for lunch in Antwerp, then having a drink in Rotterdam, and continuing to stop at halfway points on his way to Amsterdam.

The problem with this idea is that to be at a point is to be at rest at that point. Of course, someone could choose to drive to Amsterdam that way. But he or she might also drive straight through without stopping. Someone who drives straight through without stopping will never be at the halfway point, although they will pass through it. So perhaps our skeptic will say, "So he passes through the halfway point, and then through another halfway point, etc., so he never reaches his goal." But at this point the Paradox has lost its power. The Paradox depends on analyzing continuous motion as consisting of a series of discrete smaller steps. Once the idea of breaking continuous motion down into discrete steps is rejected, there is nothing to prevent the traveler from reaching his or her destination.

To refute Zeno's Paradox, it is necessary to challenge the very first statement: "First the traveler must go halfway." If this statement is allowed, then the Paradox becomes difficult to avoid. But in fact a traveler who goes from Point A to Point B does not ordinarily do so in a series of discrete steps. To go half the way to a point is to be at the halfway point. But to be at a point is to be at rest at a point. So it is not correct to say "First the traveler must go halfway."

An analogy with American football illustrates this point. When a quarterback throws the ball over the goal line to a receiver in the end zone, the ball has no problem flying through the air into the end zone because its motion is continuous. The same point could be made with international football (or soccer) when the ball is kicked into the opponent's net for a goal. Zeno's Paradox more closely resembles a series of rushing downs where the team gets ever closer but doesn't actually get in. In this case, the ball is stopped at the halfway point, then at another halfway point, etc. Of course, in football, the team has only four downs or tries to get the ball into the end zone.

To produce Zeno's Paradox, we have to imagine a football that is not three-dimensional but only exists at a dimensionless point. Then we have to imagine that the team has an infinite number of downs or attempts to get the ball into the end zone. Again, the team will either make it over the goal line or come to a point where they cannot make a further movement without crossing the goal line. But this is the practical solution again. The Paradox is only made possible by conceiving a three-dimensional object such as a football as being at a dimensionless point.

Points Lie on the Line, Not *in* the Line

Anyone who pays attention to our language will realize that we speak of points being on a line more naturally than of points being in a line, or in part of a line. The idea that points are in a line is more a result of mathematical analysis than of understanding the language of mathematics. But what is the difference between points being on a line and points being in a line?

Someone who is sitting on a fence is not part of the fence; instead, his body is physically touching the fence. But no one would think that a person sitting on a fence is part of the fence. Instead, the fence is made up of wood, steel, rocks or some other material, depending on the type of fence. Likewise, a book lying on a table is not part of the table, although the book touches the table.

Intuitively, it makes sense to say that points lie on a line. When we draw a point on a line, the line is typically there first, and we physically mark the point on top of the line. We might mark the point with an "X" (X marks the spot), with a round dot or with a little perpendicular line. However it is marked, it would be unusual to conceive of this "X", round dot or perpendicular line as somehow being a part of the line, while it is perfectly natural to think of the X, round dot or perpendicular line as lying on the line.

If points lie on the line rather than in the line, there is no need to introduce the concept of infinity to describe how many points there are. This is because a group of points lying on a line are not continuous; instead, they are simply a group of discrete points that are related by all being on the same line. It is the line that is continuous, but since discrete points are not part of a continuous line, they in themselves are not part of the continuous phenomenon.

A Line Is the Path of a Moving Point

What is the relationship between points and a line? As Aristotle says in De Anima 1:4, a line is the path of a moving point. Likewise, a plane is the path of a moving line. A point and a line then are intimately related, but not in the way Euclidean geometry thinks of them as being related. A line is somewhat like the trail of a meteor, except that when we draw a line using the point of a pencil, the line is static and remains visible.

How Many Points Lie on a Line?

If we cannot analyze a continuous line as an infinite set of arealess points, how should it be analyzed? We have already said that points lie on the line, but should not be considered part of the line. In this analysis, it makes sense to think of points as discrete units that sit on the line. If it is the number line that is being considered rather than a line that consists of the distance between two points, then these points can be considered as dimensionless in the Euclidean fashion. Euclidean mathematicians will consider these as representing an infinite number of points. In a non-Euclidean analysis, these points will have area. This area will vary with the unit of measurement that is specified. The Euclidean analysis doesn't reintroduce Zeno's Paradox with points lying on the line, as long as these are not conceived as representing the location of three-dimensional objects.

If we conceive of points as having area, no matter how small, and as sitting on the number line rather than being a part of it, then there is no possibility of introducing Zeno's Paradox. For example, consider how many points exist between 0 and 1 on the number line. If each point is defined as having a width of 1/1000 of an inch where the distance between 0 and 1 is considered to be one inch, then there are 1000 points between 0 and 1. If someone is considered to be located at one of these points, Point A (for example, point 200), and wants to go halfway to Point B, which is 1, he will go to point 600. From there, to again go halfway to Point B, he will go to point 800, then to point 900, then to point 950, and on to Point 975. In Zeno's Paradox, he would next go to point 987 1/2. However, this move is not allowed since there are no half points in this system. Zeno's Paradox is prevented if points are considered as having area or width, no matter how small. Once their area or width is defined, then it is not possible to place another point between any two points on this line.

Pursuing this non-Euclidean conception of a point as having area or width, no matter how small, it is always possible to redefine the width or area to any level of precision desired. For example, a point can be defined as one millionth of an inch, or one trillionth of an inch. Then there are a million points between 0 and 1, or a trillion points. These points are conceived as touching each other at the edges so there is no room for additional points once their size is determined. And since fractions of a point are not allowed, Zeno's Paradox cannot be generated. If a smaller or larger point is required, then the unit of measurement needs to be defined accordingly.

The Eiffel Tower All Over Again

If we think of the Paris-to-Amsterdam example, this analysis makes perfect sense. In this example, a three-dimensional car is used as the starting point to measure the driving distance in kilometers from the Eiffel Tower in Paris to the Van Gogh Museum in Amsterdam. In this case, the car traces a line as it moves along the route. At the starting point, the car is sitting on the road; it is not part of the road. As the car moves along the highway, it traces a path on the highway between the two locations. It is possible to view this path as a dimensionless line, using a Euclidean concept of a line. However, it is equally possible to view this path as being three-dimensional. If the starting point is three-dimensional, the car will trace a three-dimensional path along the road as it moves from the Eiffel Tower to the Van Gogh Museum. In this case, the car is a three-dimensional point consisting of width, height, and length. The car is used to measure the distance between the Eiffel Tower and the Van Gogh Museum by tracing a three-dimensional line along the road from the starting point to the ending point. So the idea that a line is the path of a moving point means that the car traces a three-dimensional path (line) from the Eiffel Tower to the Van Gogh Museum.

When Boundaries Matter: Defining Points and Lines

In many cases, determining area by treating the boundaries of a figure as being a line with no width works well. There is often no issue about exactly where the boundary lies. The boundary area between the end of one inch and the beginning of the next inch on a ruler is conceived of as having no width, so that the question of where to begin and stop measuring simply does not arise. The lines on the ruler are lined up with the object being measured, and an inch is marked off.

Treating a boundary line as one with no width works quite well in some cases, particularly when the boundary line is so thin relative to what it borders that no purpose would be served by treating the boundary line as having width. For example, a piece of rope that separates two tracts of land may be so thin relative to the size of the land that no purpose would be served by trying to specify the boundary more precisely. Even if there is a small portion of land that lies directly on this boundary, this portion is so small that it can be treated as nonexistent for the purposes of dividing the two tracts of land.

However, the case may be different when what lies on the boundary becomes important, or when the boundary line is large relative to the size of the marked area. For example, if gold lies on the boundary line between two properties, it may become important to try to specify which portion belongs to which property. With the center line on a highway, the line is a dividing line between the two sides of highway and doesn't belong to either side. The line is significant in size relative to the width of the road, even though it may be only several inches wide, and is an example of a boundary line with important width. Cars are not allowed to cross this line in most circumstances except to pass another car on a broken line. A doorway between two rooms provides a similar example of a boundary that has width. The area within the doorway typically doesn't belong to either room; it is there as a three-dimensional dividing line between the two rooms.

Football provides another example where the physical boundaries have width. An American football field is marked off with nine 10-yard markers. Each yard line is several inches wide. If the football rests anywhere on these lines, it is "on the 10 yard line," for example. However, at the goal line, Euclidean geometry takes over again. To score a touchdown, the player with the ball must position the ball so that it breaks the plane of the goal line before he is "down." Here the inside edge of the goal line is treated as marking a vertical plane with no thickness that the ball must break for a goal to be scored.

In baseball, the situation is similar. Chalk lines are laid from home plate down either line to distinguish fair territory from foul territory. These chalk lines are several inches wide. However, if a ball lands on the chalk line, it counts as a fair ball. It is only foul if it lands outside the chalk line. So in this case, the chalk line is treated as an extension of fair territory. The all-important "foul pole" is really a "fair pole" since balls that hit it are considered to be still in play.

Two Conceptions of Points and Lines

So does it make sense to treat lines as having width? To answer this question, let us look at the function of measurement. Measuring the area or volume of an object is typically done to find out how many units of area or units of volume it contains. When someone is baking a cake, for example, that person wants to know how many cups of flour he or she is putting into the cake. Likewise, quantities are important in commerce. A customer who buys a gallon of milk wants to know that she is getting one gallon, not some percentage of a gallon. Two functions of measurement, then, are to specify quantities for practical matters such as recipes and to ensure that people get the advertised quantities of products.

If we treat lines as having no width, this may have no practical impact in some situations. If someone wishes to divide a piece of cake into two equal slices, he or she may simply mark a line in the middle and physically divide the cake by cutting along the line. This act of division forces all particles of cake into one side or the other, and creates two pieces of cake where formerly there was one. Of course, some crumbs may result that are particles of cake that didn't stick to one piece or the other, but these are insignificant byproducts of the division process.

When the quantities are not being physically divided but only divided by a line, as in the border between two towns, the width of the line may make a difference. In some cases, where the border is disputed, a no-man's-land may be specified to mark an area between two provinces or countries that belongs to neither one. For example, the Korean Demilitarized Zone (DMZ) is a 160 mile long and 2.5 mile wide border between North and South Korea. It was created as part of the armistice agreement between North and South Korea in 1953. It is a buffer zone between the two countries, it is roughly located at the 38th parallel, and is not part of either country.

In mathematical examples, theoretical problems arise in specifying the exact border or boundary of geometrical objects. (The terms "border" and "boundary" are synonyms, except that "boundary" is often used to refer to a dividing line between two areas, including countries or tracts of land. The term "border" is often used to refer to the edge of a geometric or physical object when it is not bounded by another similar area.) It is reasonable to wonder, for example, whether the area of a circle includes only the area within the circle or whether it includes the border of the circle as well. This is especially true since the area of a circle is, by conventional mathematics, specified by an irrational number, so that clarifying exactly what the reference of the expression "area of a circle" is might shed some light on our inability to specify this area with rational numbers. The same question could be asked about rectilinear figures, although the corresponding question involving irrational areas does not arise for them.

What Is a Line?

Whether a line has width depends on what we mean by "line." Adolf Grunbaum comments on this issue. Speaking in the context of a discussion of Cantor's set theory, he says:

> No clear meaning can be assigned to the "division" of a line unless we specify whether we understand by "line" an entity like a sensed "continuous" chalk mark on the blackboard or the very differently continuous line of Cantor's theory. The "continuity" of the sensed linear expanse consists essentially in its failure to exhibit visually noticeable gaps as the eye scans it from one of its extremities to the other. There are no distinct elements in the sensed "continuum" of which the seen line can be said to be a structured aggregate." From "A Consistent Conception of the Extended Linear Continuum as an Aggregate of Unextended Elements" *Philosophy of Science* 19(4) (October 1952): pp. 288–306.

Does the idea that a line has no width make sense? This is Euclid's definition, who defined a line as a "breadthless length." (See Definition 2 in Euclid's Book One, which is quoted in Chapter Seven of this book.) This idea is also consistent with Plato's view that mathematics is about ideal, abstract objects, not about physical lines and curves. Someone who draws a rectangle and calculates the area as length × width ($l \times w$) will not retract his statement if the lines are not completely straight or if the length and width do not form an exact 90° angle. The equation is not about the physically drawn rectangle; instead it is about an abstract set of lines and relations that the drawn figure represents.

The idea that lines do not have width was Euclid's view, it was Plato's view, and it is the established view of Euclidean geometry as it is taught today. Instead of arguing against this view, which most people take for granted, I wish to present an alternative conception that proceeds from a different set of assumptions. Rather than being competing geometries, these are simply two different ways of analyzing the fundamental concepts of point, line, and area. I propose a geometry in which points have area and lines have width. I believe that this geometry more closely captures how we actually conceive of points and lines in certain circumstances, as discussed above.

The idea that a line has width follows from Aristotle's definition of a line as the path of a point in motion. The width of a line is equal to the diameter of the point used to draw the line. This does not mean we have abandoned Plato's view that geometry is about idealized objects rather than physical drawings. It only means that the abstract lines represented by physically drawn lines are conceived as having width. I propose to call this Wide Line Geometry.

Wide Line Geometry

Aristotle's definition of a line as "the path of a moving point" seems preferable to Euclid's definition of a line as a "breadthless length." Of course, this leaves the concept of a point undefined. In Chapter Seven, I devote the first three axioms of Circular Geometry to defining the concept of a point.

It is possible to treat a line as having no width for the purposes of measurement. Any drawn line, no matter how thin, has some width. The width of a drawn line is similar to the duration of a unit of time. It is not possible to specify any period of time that does not have duration. One hour, one minute, one second, one millisecond and one nanosecond all have some duration. It is simply not possible to specify a unit of time that has no duration. Of course, sometimes it is convenient for measurement purposes to consider a unit of time such as a second as being "a point in time" without duration. Likewise, for some measurement purposes it is useful to treat both points and lines as being dimensionless.

While drawn lines have width, Wide Line Geometry is not about drawn lines. Wide Line Geometry retains Plato's view that drawn mathematical points and lines are about ideal objects that are represented by these drawn points and lines, as stated above. The merit of Wide Line Geometry is that it coincides more closely with the way we actually treat lines in certain situations, as discussed above. The chalk line on the edge of a baseball field, the yard lines in an (American) football field, and the lines dividing the highway into two sides are three common examples in which we treat lines as having width. In fact, Wide Line Geometry comes much closer to capturing the way we actually conceive of points and lines than does Euclidean geometry with its "breadthless lengths."

Lines, and the Natural and Real Number Lines

Some concepts are so fundamental that it is difficult to give a meaningful definition of them in more intuitive terms. Euclid's first definition in Book One of Euclid's *Elements* is "A point is that which has no part" (see Chapter Seven). This definition could be criticized on the grounds that objects cast from a mold do not have parts, but are not considered points. For example, spoons, candlesticks, statues, pots, bullets and other objects cast from molds do not have parts. Of course, Euclid is trying to define a mathematical concept, not one that applies to physical objects, but his definition still applies to these physical objects.

Euclid defines a line as a "breadthless length," as has already been noted. This definition uses the concept of length to define what a line is, but the concepts of line and length seem very closely related. Every line has length, and Euclid makes it a matter of definition. As far as the "breadthless" part goes, this definition cannot apply to physical or drawn lines, since every drawn line has breadth, or width. Euclid here is referring to a line as an abstract mathematical concept that is represented by a physical or drawn line – an abstract line with no width.

It is also important to distinguish between a line and the number line. The term "line" is broader than the number line. People draw lines all the time in creating figures, in writing and in laying down the yard markers on a football field. The number line is an abstract concept that is used in mathematics to refer to a line with numbered points on it. According to the analysis presented here, these points lie on the line rather than in the line. Mathematicians distinguish between the natural number line and the real number line. The natural number line is a straight line with marks on it corresponding to the "counting numbers" or integers: 0, 1, 2, 3, 4, etc. Typically, the natural number line has 0 in the center with the positive integers spaced evenly to the right and the negative integers spaced evenly to the left. In traditional and Euclidean mathematics, the natural number line extends infinitely in both directions.

The real number line is similar to the natural number line, except that it also contains irrational numbers like the square root of 2 and transcendental numbers such as π. Like the natural number line, the real number line is conceived of as extending infinitely in either direction. The relation between the natural number line and the real number line is the subject of a famous hypothesis by Georg Cantor. In 1878, he formulated the hypothesis that there is no number set whose cardinality is greater than the set of natural numbers and smaller than the set of real numbers. This is known as "the continuum hypothesis." The cardinality of a set is the number of elements of the set.

Both the set of natural numbers and the set of real numbers are infinite sets. According to a proof developed by Cantor in 1873, the set of real numbers is a larger infinity than the set of natural or counting numbers. This led to the idea that infinite sets have different sizes, and contributed to the development of set theory. In 1900, David Hilbert included the continuum hypothesis as the first in his list of 23 "unsolved problems." Since that time, while minds as great as Bertrand Russell, Kurt Gödel and Paul Cohen have addressed this topic, the continuum hypothesis remains essentially unsolved today, and some have said it is undecidable because it depends on what set theory is being used.

Is the Real Number Line a Continuum?

The continuum hypothesis is so called because the real number line is viewed as a continuum. Whether the real number line is actually a continuum depends on how we define the real number line. Conceived as a line, the real number line is a continuum. However, assuming that the real number line contains numbers, and is an infinite set, then it is questionable whether it is a continuum. Previously the idea was developed that points lie on a line, not in a line. The traditional view of the real number line is that it is a continuous line containing infinitely many real numbers that lie on (or in) the line. There is nothing wrong with this idea, provided that the real number line is not identified as the set of real numbers that lie in the line. It is believed that these real numbers are dimensionless points on the line and that another point can be placed between any two points on the line. But this is equivalent to saying that all points have an area of zero but if we add infinity we somehow get to continuity. Any number when multiplied by zero is equal to zero, and zero times infinity, assuming this expression even makes sense, is also zero. So the real number line conceived as an infinite set of discrete points corresponding to the real numbers is not a continuum.

Defining a Continuum

As a starting point for our definition of a continuum, let us return to Grunbaum's comments:

> The "continuity" of the sensed linear expanse consists essentially in its failure to exhibit visually noticeable gaps as the eye scans it from one of its extremities to the other. There are no distinct elements in the sensed "continuum" of which the seen line can be said to be a structured aggregate."

Grunbaum's comments are helpful, but he attributes the continuity of a line ("sensed linear expanse") to the way in which the line is perceived by the eye. While his comments are illuminating, the continuity of a line consists in some property of the line, not merely its appearance to an observer. He also refers to it as a "structured aggregate," which makes it sound like the line is made up of discrete elements. Any definition of continuity that makes a line out to be a finite or infinite group of discrete elements that are so close together that they appear to be continuous should be rejected, for reasons developed earlier.

It is difficult and perhaps impossible to define continuity or a continuum without using a series of negatives. A continuous line is an uninterrupted linear expanse, and a continuous sound is one that occurs without interruption or without a break. Synonyms of "continuous" include ceaseless, unbroken, incessant, nonstop and uninterrupted. On the other hand, the root meaning of "continuous" is from the Latin *continuus*, the root meaning of "to hold together," and is a more positive property.

Let us go back to Aristotle's definition of a line as "the path of a point in motion." If I draw a line on a piece of paper with a pen or pencil, my motion is continuous, and it leaves a trace of ink or graphite that constitutes a line. A line is a single entity that consists of an unbroken extension between two points. Lines can be formed in nature, as in "the lines in her face," and they do not have to be formed by a moving point. The concept of a line is essentially the concept of an unbroken extension that has the shape of the path that a point with a diameter of the width of the line would make if it were moved from one end of the unbroken extension to the other. This definition allows for lines that are not actually formed by a moving point.

The number line is a line that has points on it that represent numbers. If one chooses to believe in the real numbers, including the square root of 2 and π, then the real number line has points on it that represent these irrational numbers and is infinite in length. Another view, which I favor, is that these irrational and transcendental numbers do not exist (are unreal) and that they are better represented not by gaps in the (continuous) number line, but by the absence of points where they otherwise would be. The natural or "counting" number line is as described above, a (continuous) line with points on it that represent the positive and negative integers.

Infinity and the Number Line

What about infinity? Aren't there infinitely many numbers? Along with Descartes, I prefer to say that there are indefinitely many numbers, which means there is an unlimited number of numbers, but they do not exist as a completed set, as set theory would have us believe. This means that we will never run out of numbers, but their number is not infinite. As for irrational numbers such as the square root of 2 and π, my account of Wide Line Geometry, and of Circular Geometry in the next chapter suggests that the need for these only arises out of implicitly contradictory assumptions. For example, the need for π only arises because we are attempting to determine the exact number of squares that fit into a circle. It is just as logical to say that there is no such number as to call it π and attempt to define it as a series of nonrepeating decimals that goes to infinity.

Making a Measurement Requires a Unit of Measurement and a Level of Precision

When a measurement is made, a unit of measurement is either understood or explicitly specified. If I say this stick measures 4, I have not succeeded in giving a measurement until I specify what unit of measurement I am using. I might mean 4 inches, 4 feet, 4 yards or 4 meters. Just giving a number doesn't state a measurement apart from a unit of measurement. Often the context makes this clear, though often the unit of measurement needs to be made explicit.

A second requirement for a measurement is an implicit or explicitly stated degree of precision. This may seem less obvious, since degrees of precision are not always specified, but a few examples should make this clear. When the weatherman gives the weather, it is almost always in whole degrees. If he predicts temperatures of 75° to 80°F, people are not going to expect him to predict to the tenth of a degree, like 75.5° to 80.3°F. It is clear just by using whole numbers that the forecasts are stated in terms of whole degrees and not in terms of tenths of a degree. If someone gives the distance from the earth to the sun as being 93 million miles, no one is likely to demand that he or she translate this into inches or feet, or specify how many miles plus feet or inches. 93 million miles are equivalent to 491,040,000,000 feet or 5,892,480,000,000 inches. However, given the size of the earth, the fact that it is constantly in motion and is sometimes closer to the sun than at other times, the added "precision" of feet or inches doesn't add any more accuracy to this measurement. Furthermore, it is generally understood when this number is given that it is being given in round whole numbers, and that the number of 93,000,000 miles is most likely rounded up or down from a potentially more precise number.

Here it is understood that miles are the unit of measurement. Likewise, the height of Mount Everest is variously given as 29,002 feet, 29,028 feet, 29,029 feet and 29,035 feet. The actual height reported seems to vary slightly. In 1999, an American survey used a Global Positioning System (GPS) to arrive at a figure of 29,035 feet, with an acknowledged margin of error of 6.5 feet. While a few reports did use fractions of feet, there were apparently no published reports giving this height in inches. Feet (or meters) to the whole number is the generally accepted unit of measurement for the height of Mount Everest.

Zeno's Paradox Requires Constantly Shifting the Level of Precision

Zeno's Paradox as stated violates the idea that every measurement is made to some degree of precision. It is possible to state this Paradox in feet without shifting units of measurement. But it cannot be stated without constantly shifting the degree of precision required in the measurement. Zeno begins by innocently imagining a traveler moving from Point A to Point B. Let us suppose this distance is 100 feet. So in this example, the unit of measurement is feet and the degree of precision is a whole number of feet. Then the traveler goes halfway there, and is 50 feet from the finish line. Then he goes again, and is 25 feet from the finish line. So far, the Paradox is using whole numbers. Next he has to go 12.5 feet, so now the degree of precision has advanced to one decimal point after the whole number. Next he is at 6.25 feet, so the degree of precision has advanced to two decimal points. Next he is at 3.125 feet, then 1.5625 feet. The next step is 0.78125 feet, which is five decimal points. After several more divisions we are at 0.09765625 feet, which 8 decimal points of precision. And so on to infinity but never reaching the goal.

If the precision for a measurement is not adequate, it is always possible to specify a greater degree of precision. For example, in measuring the length of a room, it might be advantageous to state the length in feet to two digits of precision rather than in whole feet. This may also apply to measuring furniture. Once the level of precision is determined, then a measurement can be made. But no matter what level of precision Zeno specifies, he will eventually reach the limit. Any measurements made beyond that point will be rounded to the nearest whole number.

The Degree of Precision Required Varies with the Measurement

Whatever the degree of precision, there will always be some motion or length that is beyond this degree of precision. For example, the earth is considered to be 93 million miles from the sun, even though it is in constant motion. Measuring this distance more closely requires greater precision. Zeno's Paradox would have us bring out a new yardstick with greater precision every time we need to measure a distance that is beyond the specified degree of precision. This amounts to constantly increasing the degree of precision to capture the motion that is not being captured by the current degree of precision. But this does not show that motion is impossible, as Zeno would have us believe. Instead, it shows that whenever a measurement is made, a more precise measurement is possible using a more precise instrument. As we have emphasized, the unit of measurement selected should be appropriate for the object being measured and for its length.

Precision in time is also important in a parallel way. If I ask what time it is and you say "It's noon," I am not likely to object if it is only two minutes before noon. However, sometimes we need to know exactly to the minute or even to the second what the time is. This precision is often critical in making appointments on time and in sporting events. Even so, the measurement is not usually carried out beyond the precision of seconds except in scientific measurements and in certain sporting events. An example is the last minutes of many professional basketball games, when time is measured in tenths of a second.

There are 1,440 minutes in a day and 86,400 seconds. This doesn't vary, except in other systems such as decimal time, but it is important to note that, just as points have area, seconds have duration. It is easy to think of a second as a "point" in time, but a second lasts a second, a minute a minute, and an hour an hour. There is no such thing as a unit of time without duration. Otherwise, there would be infinitely many points in time, which there are not.

Most people have experienced an hour or even a minute that seems like an eternity. I experienced such a period while waiting for my Ph.D. dissertation committee to come back with their verdict after my oral exam (fortunately, it was positive, or I probably wouldn't be writing this book). While we use the expression "point in time," units of time are very much like points on a line. Just as units of time have duration, so points on a line have area. And the amount of area varies with the unit of measurement.

Applications to Flow and Process Measurement: How Long Is the South Caucasus Pipeline?

Length plays a critical role in flow measurement and elsewhere in the process industries. The most common units of length in the American system are the inch, foot and mile. The yard may also play a role, though it is more common to use feet than yards in flow and process measurement contexts. In the metric system, the common units are the millimeter, centimeter, meter and kilometer.

An example of the use of length in the natural gas industry is the length of the South Caucasus Pipeline. This pipeline connects natural gas produced in the Caspian Sea to Europe via Georgia and Turkey. The pipeline is nearly 435 miles or 700 kilometers long. The Nord Stream Pipeline connects Russia with Europe via the Baltic Sea. Nord Stream is 760 miles or 1220 kilometers in length. The South Stream natural gas pipeline connects Russia with Europe by means of the Black Sea. South Stream is 560 miles or 900 kilometers long.

Here the units of length are either miles or kilometers. But how precise are these measurements? When measuring the length of a pipeline, there is some leeway in terms of the starting point and ending point. This is somewhat like the Paris-Amsterdam example. As long as a prospective traveler is somewhere in Paris, he or she can use that position as a starting point. Likewise, the starting point of a pipeline that is connected to a natural gas well could be considered somewhere in the vicinity of the well. There will also be some flexibility in the endpoint since it could be somewhere near the point of delivery of the natural gas. Since these measurements are made in miles or kilometers, it is reasonable to assume that the degree of precision in this measurement is somewhat limited, unless it becomes important to make a more precise measurement.

Length in Flow Measurement: Does a Pipe Circumference Have Width?

As we have stated, the concept of length is critical in flow measurement. Fluids typically travel in round pipes, and flowrate is measured by the classic equation:

$$Q = V \times A$$

Here Q is flowrate, V is velocity and A is cross-sectional area. The area of a pipe is typically determined by the equation πr^2. Here r is the radius, which is one-half of the diameter of the pipe. The diameter of a pipe is a straight line that runs from one side of the pipe to the other through the center of the pipe. Radius is a measurement of length, and is fundamental to measuring flowrate.

It is conventional to distinguish between the inside diameter (ID) of a pipe and its outside diameter (OD). When measuring flow, it is the inside diameter of a pipe that is taken into consideration. While it is common in Euclidean geometry to think of the circumference of a circle as having no width, this is not how pipes are in reality. While they are indeed round, for the most part, their circumference has width (the wall thickness), and this is important for some types of flow measurement, such as clamp-on ultrasonic flowmeters.

One common problem in determining an accurate flowrate is that buildup can occur on the inside of pipes. This impacts the accuracy of flowrate measurement, since it decreases the inside diameter of the pipe. For example, the accuracy of ultrasonic flowmeters that send an ultrasonic signal across the flowstream and back can be affected if pipe buildup occurs, since it reduces the length that the signal travels. Clamp-on ultrasonic meters that send an ultrasonic signal through the pipe wall can also be negatively impacted by pipe buildup. The signal may already be attenuated by the pipe wall, and pipe buildup can further attenuate the signal.

ANSI, ASME and DIN Flanges: Challenges in Universal Length Measurement

Another issue involving length measurement has to do with American and European conventions of measurement. A flange is a ring or collar, usually with holes for bolts or screws, that is attached to an object such as a transmitter or flowmeter to allow it to be attached to another object or to a pipe. The American Society of Mechanical Engineers (AMSE) and the American National Standards Institute (ANSI) have issued guidelines that all American flange manufacturers can follow. In Europe, the Deutsches Institut für Normung e.V. (DIN) (German Institute of Standardization) issues parallel guidelines for the construction of flanges.

All would be harmonious except that the ANSI and ASME standards are issued in terms of inches and the DIN (German) standards use the metric system. While one inch is treated in many contexts as being equivalent to 25 mm, in reality one inch equals 25.4 mm. So the American and metric systems are similar but are fundamentally incompatible for constructing devices that need to be threaded together. The differences become more pronounced as the pipe sizes get larger. The Japanese Industrial Standard (JIS) offers still another set of standards.

The solution from many flowmeter and transmitter manufacturers, especially in Europe, is to offer flanges and connectors conforming to both ANSI/ASME and DIN standards. Flowmeters and transmitters are not commodity items, and they are typically ordered by specifying a wide range of parameters including materials of construction, pipe size, software, display type and many others. As a result, end users can specify whether they want flanges or connectors conforming to ANSI/ASME or DIN standards. This allows European companies to sell both into Europe and also into the Americas.

It should be clear that units of length play a fundamental role not only in our daily lives, but also in the more specialized areas of flow measurement and process control. The units of length used today, like those of time, are in many cases rooted in ancient times. As has happened with the development of advanced methods of measuring time, technology has enabled us to develop more precise and accurate methods of measuring length. For example, the length of the standard meter is now measured in light waves rather than the physical length of an iron bar. One of the most confusing issues remaining, however, is differing measuring systems, especially the difference between the American system and the metric system. These topics could perhaps be more aggressively addressed by the scientific community, but there is no single entity with sufficient authority to either mediate differences between the two or to generate a wider adoption of one system over the other. Between the two, however, the metric system is by far the more widely used around the world.

The discussion of Zeno's Paradox may seem somewhat like the proverbial discussion of how many angels can dance on the head of a pin. However, this discussion reveals vitally important questions about our most treasured and commonly accepted mathematical assumptions. It also relates to questions that undermine the validity of calculus, although there was no space to fully explore those topics here. What hopefully will be apparent from the above discussion is that there are alternative ways to conceive geometry, or that there are alternative geometries, and that just because mathematical beliefs have been held for 2,500 years does not guarantee that they are true or even that they are coherent. I view Wide Line Geometry as a viable alternative to traditional and Euclidean geometry, and believe that it coheres much more closely to our intuitions about boundaries and lines than does traditional and Euclidean geometry.

Units of Length: One Furlong and a Doorway

The following are some common units of length measurement that you may encounter in a variety of contexts:

Egyptian cubit – The length of the arm from the elbow to the extended finger (about 3000 B.C.)

Ancient Greek foot – The width of 16 fingers

Ancient Roman inch – The Greek foot divided into 12 sections (the unicae)

Yard – Decreed by King Henry I in the 12th century who ordered a standard yard to be constructed with a length of three feet.

The distance between the end of your thumb and the first joint is about one inch.

The width of a standard doorway is one yard.

1 meter = 3.28 feet

1 inch = 25.4 millimeters

1 foot = 0.3048 meters

1 mile = 5,280 feet

1 mile = 1,760 yards

1 mile = 1.609344 kilometers

1 kilometer = 0.62137 miles

1 yard = 914.4 millimeters

1 yard = 0.9144 meters

1 nautical mile = 1.150779 miles

1 furlong = 660 feet

1 furlong = 7920 inches

1 centimeter = 0.3937 inches

1 inch = 2.54 centimeters

1 centimeter = 10 millimeters

1 meter = 100 centimeters

1 kilometer = 1000 meters

Morley's Point: Length

I learned a lot reading Jesse's chapter on length. It forced me to think about length as a dimensional and language concept. Length is defined by the observer; it is not defined by engineering, and the definition varies depending upon the application. The observer can be a person, mathematics, simulation, or a process such as liquid flow. Each observation tends to modify the meaning of the term "length."

When you want to know something, ask a child. A child's observation is simple and straightforward. My secretary's grandson, 10-year-old Jaden, said that it is "how long something is." Length is usually attached to two other concepts – height and width. Each of them has a smaller number, whereas length is dimensionally constrained.

What happens if I cut a long cubic object shorter than the length? Now what is the length? Sometimes the length is just the longest line in a cubic representation. Sometimes it is from corner to corner. When we buy a 32-inch television, that's the corner-to-corner visible screen. The length is related to the contextual statement being made by the observer.

Length is also related to time. The length of time it takes to get to the airport is an observable variable. Is it just the time on the road or is it from your door to the airplane gate?

Around 480 B.C., Zeno of Elea made some discussions about length. A modern version of his conjecture might look like this: If you are 100 miles away from the city of Boston, Massachusetts, you can approach Boston by using the distance as the speed of the vehicle. At 100 miles, you can go 100 mph; at 50 miles, 50 mph and on down to 1 mph at one mile away. What is the length of the trip? Philosophical amateurs cannot answer that question.

Some solid-state people use length as the longest line you can draw inside a solid object. The measurement I like the best is the measurement of the Harvard bridge, which is measured in smoots. In 1958, an MIT frat student, Oliver Smoot, was used as the standard. The bridge length is 364.4 smoots (620.1 m) plus or minus one ear. Even when they repair the bridge, they keep the markings. You can see them still today.

There are also the variables. What is the length of a rubber band? Or the length of the coastline of England? The length of the coastline, if measured by the mile, is shorter than if measured by feet. If you make a measurement device shorter and shorter, the apparent length gets longer and longer, and this keeps going all the way down. For example, in a pipeline that runs hypothetically from St. Petersburg to Moscow, we measure "as the crow flies." But the pipe itself is actually longer than that. It has expansion joints and corners, and has the same "mistakes" as the coast of England. When designing systems, we have to make sure the view of the length of the pipe is from the material's view, not the observer's view.

As a machinist (what I did to help earn my tuition in the early 1950s), I felt that the numerical control (NC) machine had to know and work with the definition of the machining process, not as an intellectual argument. We should remember that the "person" who views length is the flow going through the pipe or the computer numerical control (CNC) and not the human watching it.

Some other personal observations I've made and the results of this research examine the social aspects of length. While waiting for a movie to open and suddenly all the seats fill up, the length of wait becomes infinity (or a couple of hours), so length can be dynamically nonlinear.

Sometimes length is artistically significant. I remember my noble wife,

Shirley, insisting that when she hung the wash on the clothesline (pre-electric dryer) she would arrange them according to the length of the clothing. As an inexperienced husband I tried an experiment. I moved one of the clothing elements on this line and exchanged it with another. I slept on the sofa.

My friends who wear leather elbows argue forever about what is length. Is it cross-dimensional? Does length really matter? It is the observer who defines length. Live with it.

Mathematical discoveries, like springtime violets in the woods, have their season which no human can hasten or retard.
– Johann Carl Friedrich Gauss

Chapter Seven

Going in Circles and Toeing the Line to Measure Area

Johann Carl Friedrich Gauss

This chapter looks at the concept of area, specifically the areas of circles and squares, which are the building blocks of area measurement. We explore some conundrums—such as how to count borders when calculating area and how to reconcile traditional geometry, which is uncompromisingly linear, with a world that is full of waves, curves, circles and other nonlinear shapes.

The concept of area builds on notions discussed in earlier chapters, such as point and length, and the need to specify a unit of measurement when measuring length. It was argued that the fundamental units of measurement have dimension, and that contradictions result if they are assumed to have zero area. And the same idea applies to time. Even though, for convenience, we may treat the idea of a second or an instant as having no duration, the fact remains that even the tiniest fraction of a second has some duration, however brief, and therefore has a kind of "area."

Area – aided by its servants, length and width – plays a critical role in our lives. Measurement of land by area, for example, dates back more than five centuries. The term "acre" is derived from the Middle English term "aker," and it originally referred to the amount of land that someone could plow in one day. Today it is equivalent to 43,560 square feet.

While acres are used to measure land, the concept of area is important in many other ways too. We may calculate the area of a desk, a room, a driveway, a yard or a kitchen cabinet. The concept of area is fundamental to commerce and to the transfer of ownership of property from one person to another. When the property is in a pipe and is a fluid with commercial value, this is called *custody transfer* and is based on the use of area to calculate flowrate.

The concept of volume, not explored in this chapter, is perhaps the rich city cousin to area. Area is two-dimensional, while volume is three-dimensional. Like area, volume is critical to our daily lives. In the American system, volume is measured in cups, pints, quarts and gallons. In the metric system, it is measured in milliliters and liters. While there is not an exact equivalence between these units, there is an approximate equivalence: one liter is equivalent to 1.567 quarts. Milk is measured in quarts and gallons, while liquids, like syrup and honey, are measured in pints and quarts. Many liquids, such as water, are measured in fluid ounces and in milliliters, but they are sold in pint-sized, quart-sized, or gallon-sized bottles or plastic containers. There is a move in the United States to implement the metric system, so many liquids are measured in both fluid ounces and milliliters. (The same dual marking of units on packaging applies to solids such as flour and sugar, which are measured in cups or are weighed in ounces and pounds and also in grams or kilograms.)

Area: Typically Defined in Square Units

Area can be defined as the amount of space a two-dimensional figure or object takes up. For area, it is natural to think of a plane figure such as a square. However, the area of the surface of a cube and the area of the surface of a sphere also exist, although this is also two-dimensional. The area of the surface of a cube is the sum of the areas of its six sides. When area becomes three-dimensional, it is called *volume.* There are various formulas for determining area, depending on the figure or object's shape. These formulas reveal the amount of area based on some unit of measurement. One of the most familiar ways to calculate area, used for squares and rectangles, is to multiply length times width, so the area of a rectangle that is 3 inches by 4 inches is 12 square inches.

Area is typically calculated in terms of square units, such as square inches, square feet, square meters, square miles etc. In calculating area, each side is bounded by a line with no width, and the area is given in terms of the number of square units the figure or object occupies. With physical objects such as a tract of land or the surface of a desk, the length and width are represented by geometric lines that go in a straight line along the edges.

Euclidean Geometry

Traditional and Euclidean geometry is based on definitions postulated by Euclid in Euclid's *Elements*, which was published in 300 B.C. The first five definitions are worth looking at here, particularly since the axioms of Circular Geometry, stated later in this chapter, are modeled to some extent on Euclid's axioms. Euclid's first five definitions (also called *axioms*) are as follows:

1. A point is that which has no part.
2. A line is a breadthless length.
3. The extremities of a line are points.
4. A straight line is a line which lies evenly with the points on itself.
5. A surface is that which has length and breadth only.

Source: *Euclid's Elements, Book One*, Cambridge: Cambridge University Press, 1905.

Book One contains many additional definitions, along with postulates and axioms. It includes the famous parallel "fifth postulate," the denial of which has formed the basis for several different "non-Euclidean geometries."

While Euclid recognized the existence of circles and circular area, his analysis of this area is strictly based on straight lines. In fact, as is shown in Figure 7-1, the unit of measurement for measuring circular area is the area of a square. The need for π in the formula for the area of a circle arose because no rational number of squares can be made to fit inside a circle. This idea is explored in more detail in the following section.

Measuring the Area of a Circle – Trying to Fit a Square Peg into a Round Hole

If you studied mathematics or geometry in high school or college, you probably learned the following formula for the area of a circle:

$$A = \pi \times r^2$$

Here A is the area of the circle while r is the radius of the circle. The number π, which represents the ratio of the circumference to the diameter of the circle, is an irrational number that has never been completely specified.

What is this formula actually asking us to do? If we look at the geometry of this formula, it looks like the diagram in Figure 7-1.

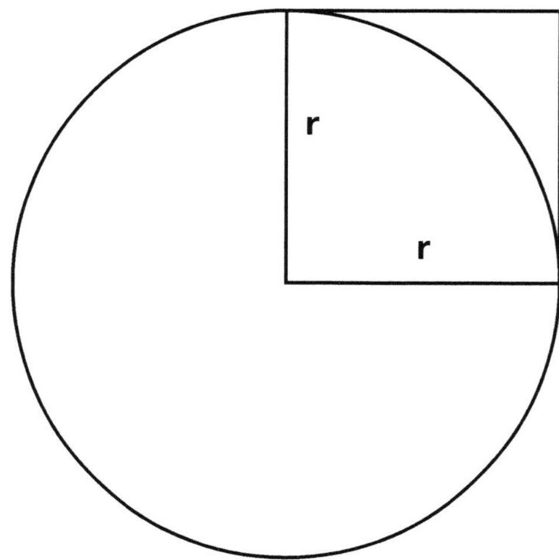

Figure 7-1. The Square as the Unit of Measure for Circular Area.

The value r^2 gives the geometric area of the square in the above diagram. The formula for the area of a circle, then, tells us that π squares with sides equal to radius r fit into the area of a circle with radius r.

What, if anything, is the problem with this formula? And why do we need to have π in the formula? The reason for π is that there is no rational number of times that a square can fit inside a circle. Since the formula for circular area, $\pi \times (r \times r)$, involves the area of a square, (r × r), this formula actually involves calculating how many squares will fit into a circle. It is often said "You can't fit a square peg into a round hole." This common saying reflects the insight that the area of a square cannot be used as a unit of measurement for circular area. Since there is no definite number of times that a square will fit inside a circle, the value π has to be included to create a usable formula for circular area. In fact, the number is π, a nonrepeating, irrational number that mathematicians to this day cannot fully define.

The relationship between circular area and square area is that they are incommensurable. What this means is that they cannot both be measured exactly using the same standard or unit of measure. Straight lines and squares work fine for squares and rectangles, but they do not allow us to provide exact values for the areas of circles.

In order to more accurately measure circular area, we may need to consider using a new geometry.

The Trouble with Euclidean-Cartesian Geometry

Traditional geometry rests on two main pillars: the Euclidean axioms that create the conceptual underpinning of geometry and the Cartesian coordinate system that provides an *x* and *y* axis (and a *z* axis, in three-dimensional geometry) in terms of which, points can be located.

One issue with Euclidean geometry consists in the Euclidean conception of the point and the relationship between points and lines. In Euclidean geometry, a line is made up of infinitely many points, each of which has zero area. No matter how many times zeros are added together, however, a positive value never results. And adding infinity into the equation doesn't change anything, since zero times infinity is still zero. While many attempts have been made to explain away this anomaly, it still remains unexplained and unexplainable.

There is also a problem with the Cartesian coordinate system. While the straight-line framework of the Cartesian coordinate system works well for analyzing the areas of squares and rectangles, it is less successful as a frame of reference for curved and circular areas. Finding the area of a circle requires the use of π, which as we have seen is the ratio of the circumference of a circle to its diameter.

Why π?

The need for π arises because square or rectangular area and circular area are incommensurable. This means they cannot both be analyzed from the same perspective, or point of view. Instead, to accurately describe and analyze circular area it is necessary to adopt a frame of reference that is appropriate to circular area.

It is time to reconsider certain elements of Euclidean-Cartesian geometry. The belief that a line is composed of infinitely many points with no area leads to contradictions. Likewise, the idea that it is possible to accurately represent circular or even curved area in terms of the straight-line frame of reference supplied by the Cartesian coordinate system does not appear to withstand scrutiny.

If the areas of squares and circles cannot both be measured exactly using the same unit of measurement, we have a choice. One is to continue as we have been, using square areas as the unit of measurement for circular area. This has the advantage of familiarity, provided we don't mind using π. A second alternative is to use a different unit of measurement for circular area. This is the alternative I would like to suggest here.

Dividing a Circle into Four Equal Areas by Inscribing Two Smaller Circles

In order to define an alternative unit of measurement for circular area, it is worth examining several important relationships that exist when a circle is inscribed with two circles of equal area. In this example, the circumference of the two smaller circles touches on the circumference of the large circle and also passes through the center of the large circle. I will examine these relationships in terms of traditional and Euclidean definitions of the area of a circle. This circle with its inscribed smaller circles is shown in Figure 7-2.

An Alternate Unit of Measure for Circular Areas: The Round Inch

Instead of defining circular area in terms of square inches, I propose to define it in terms of round inches. A round inch is the area of a circle with a diameter of 1 inch.

If we start with the basic concept of the round inch in place of the square inch, it is possible to represent circular area in terms of how many round inches fit into a circle. It is then possible to eliminate π from the calculation and substitute in its place a different formula that has no need of π. I am proposing to call this new geometry Circular Geometry.

If we use the round inch as the unit of measurement for circular area, this unit of measurement looks like the drawing in Figure 7-2.

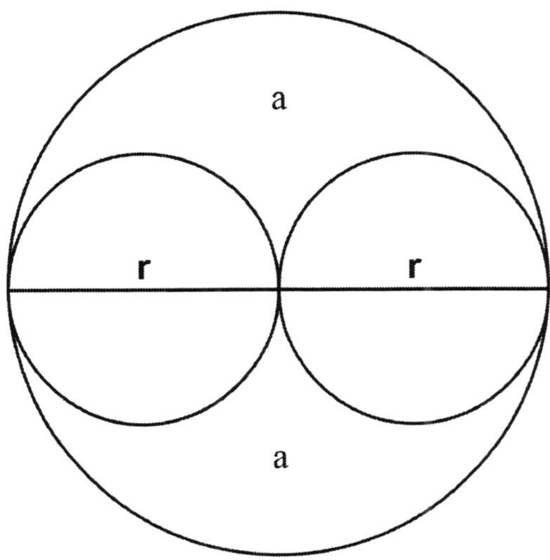

Figure 7-2. Dividing a Circle into Four Equal Areas

In Figure 7.2, the area of each of the two smaller circles is equal to A/4, where A = the area of the larger circle. This is shown as follows. In Figure 7-2, the radius r of the large circle is 1 inch. Using the traditional equation for the area of a circle:

$$A = \pi \times r^2$$

it follows that the area A of the large circle equals $\pi(1^2)$:

$$\text{Large Circle Area } A = \pi \times (1^2)$$

Since 1^2 equals 1, the area of the large circle equals $\pi \times (1)$:

$$\text{Large Circle Area } A = \pi \times 1$$

Since $\pi \times 1 = \pi$, the area of the large circle = π:

$$\text{Large Circle Area } A = \pi$$

Each small circle has a diameter of 1 inch (r = 1 inch, where r = the radius of the large circle), and so has a radius of 1/2 inch. Using the formula for the area of a circle, the area of each of the two small circles is $\pi \times (1/2)^2$

$$\text{Small Circle Area} = \pi \times (1/2)^2$$

Since $(1/2)^2 = (1/2^2) = (1/4)$, the area of the small circle = $\pi \times (1/4)$:

$$\text{Small Circle Area} = \pi \times (1/4)$$

Since $\pi \times (1/4) = \pi/4$, the area of each of the smaller circles is equal to $\pi/4$:

$$\text{Small Circle Area} = \pi/4$$

So the area of each of the smaller circles is ¼ the area of the large circle.

It also follows that the areas of the two smaller portions of the circle that are left after the two equal circles are inscribed inside the larger circle each have areas of $\pi/4$. This is because each of the circles has an area of $\pi/4$, or $1/4\pi$, so the two circles together have an area of $1/2\pi$. But the two remaining segments are equal in size and together equal $1/2\pi$. So each remaining segment is $1/4\pi$, or $\pi/4$, which means that the area of each of the remaining segments is equal to the area of each of the two inscribed circles. So inscribing these two equal sized circles inside a larger segment in the way illustrated in Figure 7-2 divides the area of the larger circle whose area is A into four equal areas. The areas of these two remaining segments are marked "a" in Figure 7-2.

Circular Mils

A similar approach to this already exists for measuring the area of round wire. In order to avoid using decimals, the area of round wire is often measured in circular mils. The area of a circular mil is $(\text{diameter})^2$, so the formula is as follows:

$$A = (\text{diameter})^2$$

A mil is equal to 1/1000 of an inch (0.001 inch). A circle that has one mil as its diameter has an area of $1^2 = 1$ circular mil. A circle that has a diameter of 4 mils has an area of $4^2 = 16$ circular mils.

What is the relationship between a round inch and circular mils? A circle with an area of 1 round inch has a diameter of 1 inch. Since a mil is 1/1000 of an inch, a circle with an area of 1 round inch has a diameter of 1000 mils. This means that a round inch is equivalent to an area of 1000^2 circular mils, or 1,000,000 circular mils. So any area that can be measured in round inches can be measured in circular mils, and vice versa.

Just as circular mils are used to measure the area of wire, round inches can generally be substituted for square inches in geometry when calculating the area of circles. Of course, just as there is no exact way to measure circular areas in terms of straight lines, there is no way to exactly measure the area of squares and rectangles using circular geometry. Just as a hammer is used for nails and a screwdriver is used for screws, so each type of geometric structure requires its own geometry.

Someone might object that the round inch is defined as equal to $\pi/4$, so Circular Geometry does not get rid of π; it actually contains π in its definition. This is obscured by taking the area of the round inch as primitive.

It is true that on a traditional and Euclidean analysis of the area of a round inch, it is equal to $\pi/4$. However, while the area of a round inch is equal to the area analyzed as $\pi/4$ on a traditional and Euclidean analysis, it doesn't follow that Circular Geometry is committed to using π, or that π is somehow implicit in the definition of a round inch. Euclidean geometry uses a square inch as the fundamental unit for describing area, and in terms of the square inch the area of a circle can only be described by using the irrational number π. By taking the round inch as the fundamental unit of measurement, Circular Geometry avoids the use of π, and can provide rational values for the areas of circles.

The Development of Non-Euclidean Geometries

While Euclidean geometry remained largely unchallenged for many centuries, in the 19th century several mathematicians developed alternate geometries that were based on modifying one or more of Euclid's first five axioms. Two such mathematicians include Bernhard Riemann and Johann Carl Friedrich Gauss, who were instrumental in developing several alternate geometries that modified Euclid's fifth axiom.

According to Euclid's fifth axiom or "postulate," through a point that is not on a given line there is only one line parallel to the given line. In spherical geometry, there is no line through a point not on the line but parallel to the line. In hyperbolic geometry, there are at least two distinct lines that pass through the point and are parallel to the given line. These geometries accept Euclid's other four postulates, or modify them slightly.

The following list of axioms of Circular Geometry also constitutes a non-Euclidean approach to geometry. However, these axioms are less tied to Euclid's first five axioms than are other non-Euclidean geometries. Instead, they attempt to set forth a geometry that is appropriate for mathematically describing and calculating circular area.

The Axioms of Circular Geometry

Like Euclid's geometry, Circular Geometry can be formulated as a series of axioms and definitions, formulated as follows:

1. A point is the smallest allowable unit of measurement within a system of measurement.
2. Every point has area.
3. A point is a circular figure that is considered to be indivisible for the measurement being made.
4. A line is the path of a point in motion.
5. Every line has width. The width of a line is equal to the diameter of the point being used for a particular measurement.
6. Every measurement involving points and lines is relative to a system of measurement in which the reference of the terms "point" and "line" is fixed for the purposes of that measurement. The reference of "point" determines the degree of precision used in the measurement.
7. In the Coordinate System of Circular Geometry, there is an *x* axis consisting of unit circles laid end to end in an east to west direction, each with a value of one round inch. Likewise, there is a *y* axis consisting of unit circles laid end to end

in a north to south direction, each with a value of one round inch. The point of intersection of these circles creates the integers 0, 1, 2, 3, etc.

8. To find the area of a circle, which is the number of round inches in the circle, use the formula 4 × r × r, where r is the radius of the circle. Alternatively, use d × d, where d is the diameter of the circle.

9. A point lies on the line, not in the line. Finitely many points lie on any given line.

10. A circle is a continuous closed geometric figure that is the path generated by rotating a point around a fixed point of the same size until it intersects its starting position.

11. Every circle has an inside diameter (ID) and an outside diameter (OD). The inside diameter begins and ends at the inside edge of the circle. The outside diameter begins and ends at the outside edge of the circle. The inside area is calculated by taking d to be the inside diameter in the formula d × d. The outside area includes the circle boundary, and is calculated by taking d' as the outside diameter in the formula d' × d'. The width of the boundary of a circle is equal to the diameter of the point that is used to generate the circle.

12. A plane is the path generated by sweeping a line through space. Just as a line has width, a plane has a depth equal to the width of the line.

Circular Geometry, Euclidean Geometry, and Other Geometries

The above discussion should not be construed as a refutation of Euclidean geometry. I have, however, pointed out that accepting the assumptions of Euclidean geometry does lead to Zeno's Paradox. Euclidean geometry also requires accepting the idea of a number line consisting of infinitely many arealess points. Some mathematicians have no problem with the concept of infinity. I believe the term "infinity" is a lot easier to use than to coherently define; after all, an infinite process can never be completed. I prefer to follow Descartes and use the term "indefinite," meaning "unlimited," where others may choose to use the term "infinite."

I am proposing Circular Geometry as an alternative geometry that proceeds from different assumptions, much as do other non-Euclidean geometries. However, Circular Geometry has nothing to do with Euclid's fifth postulate, as do those developed by Riemann and Gauss. Instead, it proceeds from a different definition of point and line, and proposes a circular unit of measure to replace the square. Unfortunately, it does not appear that Circular Geometry will be any more successful in providing rational values for square or rectangular area than Euclidean geometry has in providing rational values for circular area. Reality does not always conform to our wishes, and as the vain attempts to "square the circle" have shown, circular area and square areas appear to be incommensurable, in the sense that they cannot be rationally measured using the same unit of measure. Nonetheless, Circular Geometry may have some advantages when dealing with circular figures.

Circular Geometry Applications Abound – from Architecture to Flow Measurement

What are the applications for Circular Geometry? Circular Geometry can be used anywhere someone wants to measure circular area. It is true that many buildings and other structures are either square or rectangular. This is an example of geometry influencing architecture. Because most of the geometry that is taught in schools is so uncompromisingly linear (meaning that it is based on straight lines, squares and rectangles) many of the buildings and other structures that are created using this geometry reflect its underlying linear nature. On the other hand, if we look in nature, we find a wide assortment of waves, curves, circles and other nonlinear geometric shapes. The world is round, even though it looks flat, and many natural shapes are nonlinear as well. Circular Geometry is well-suited to analyzing the geometry of these structures. In architecture, many round houses and other round structures have been built, and Circular Geometry has application to these structures.

Rather than expanding upon Circular Geometry, I will discuss the implications of Circular Geometry for flow measurement. Does this new Circular Geometry have any implications for flow measurement? I believe that the answer is a resounding "Yes!"

Circular Geometry for Flow Measurement

Circular geometry is particularly useful for flow measurement. Pipes are round, and it is often necessary to determine the area of a pipe in order to determine volumetric flowrate. The formula that is usually used is the following one:

$$Q = A \times V$$

In the above formula, Q is equal to volumetric flowrate, A is the cross-sectional area of the pipe and V is the average velocity of the fluid. It is in providing the cross-sectional area of a pipe that the value of π is used in calculating flowrate. If this area is provided in round inches rather than square inches, flowrate can be calculated without the use of π.

Here are three areas in which Circular Geometry has the potential to improve flowrate calculations:

The Fundamental Unit of Flow Measurement

The first area has to do with the fundamental units of flow measurement. The fluid dynamics section of a college physics text describes two approaches to measuring flow:

> "One way of describing the motion of a fluid is to divide it into infinitesimal volume elements, which we may call fluid particles, and to follow the motion of each of these particles. This is a formidable task. We could give coordinates x, y, z to each such fluid particle and then specify these as functions of the time t and the initial position of the particle x0, y0, and z0. This procedure is a direct generalization of the concepts of particle mechanics developed by Joseph Louis Lagrange." D. Halliday and Robert Resnick *Fundamentals of Physics*, 2nd Edition, New York: John Wiley & Sons, p. 277, 1981.

A second approach was taken by Leonhard Euler, which Halliday and Resnick describe as follows:

> "In it we give up the attempt to specify the history of each fluid particle and instead specify the density and velocity of the fluid at each point in space at each instant of time. This is the method we shall follow here. We describe the

motion of the fluid by specifying the density ρ(x, y, z, t) and the velocity v(x, y, z, t) at the point (x, y, z, t). We focus attention on what is happening at a particular point in space at a particular time, rather than on what is happening to a particular fluid particle. Any quantity used in describing the state of the fluid, for example, the pressure p, will have a definite value at each point in space and at each instant of time. Although this description of fluid motion focuses attention on a point in space rather than on a fluid particle, we cannot avoid following the fluid particles themselves, at least for short time intervals dt. For it is the particles, after all, and not the space points, to which the laws of mechanics apply." D. Halliday and Robert Resnick *Fundamentals of Physics*, 2nd Edition New York: John Wiley & Sons, p. 277, 1981.

In Circular Geometry, flow is not measured in infinitesimal volume elements, but in small "flow units" (points) whose size would vary (or could vary) with the fluid being measured. These "flow units" would be defined in terms of circular area, rather than square area. It is like the particle approach, except that the volume elements are not infinitesimal, but instead are finite and definable. Then the quantity of flow is given in terms of how many of the finite flow elements travel past a given location in a given length of time. A parallel intuition would be to measure flow in drops; for example, to say, "This fluid is flowing at 1,596 drops per minute." A conversion could also be created from drops (or points) to gallons or liters.

More specifically, the "flow units" I am proposing are the points of Circular Geometry laid out in the above axioms. These points could either be defined in terms of volume or of mass. Fluid flow through a pipe could then be described in terms of how many of these "flow units" or points pass a given location in a period of time.

Application to the Flow Equation

A second area is applying this geometry to the flow equation:

$$Q = A \times v$$

Here Q = flow, A = area, and v = velocity. Area can now be measured in round inches instead of in square inches. This eliminates π from the equation, and does away with a nagging source of inaccuracy and uncertainty in flow calculations.

Inside Diameter and Outside Diameter

A third area of application is that, in this geometry, a circle has an inside diameter and an outside diameter. This inside/outside diameter corresponds naturally to pipes, which also have an inside and outside diameter. If we include Wide Line Geometry with Circular Geometry, a pipe becomes an extended circular structure with a "wide" or three-dimensional boundary. When calculating flow, it is the inside diameter (ID) that is used. The outside diameter (OD) gives the actual size of the pipe. So if it is a 14 inch pipe in terms of its outside diameter and the pipe wall is one inch thick, the inside diameter (ID) is 13 inches. In this way the pipe is viewed as a single structure with a one inch thick wall rather than as a structure with dimensionless inside and outside diameters.

Circular geometry, then, has implications for flow measurement in terms of the fundamental unit of flow measurement, in terms of the flow equation $Q = A \times v$, and in terms of measuring the inside and outside areas of pipes. Once round inches become accepted and understood as a unit of measurement, round inches can replace $\pi/4$ in the flow equation. The element of uncertainty implicit in the use of π will be a thing of the past, and flowrate can be reported without relying on the irrational number π.

Time for a Fresh Look

It is time for a fresh look at the assumptions of Euclidean-Cartesian geometry. While much work remains to be done, I believe that Circular Geometry will prove to be a viable alternative to Euclidean and Cartesian geometry for calculations involving circles. Change is often good, and in this case, change is needed. Not only can Circular Geometry put geometry on a more sound footing, it can also improve our methods of measuring fluid flow. Once the round inch is accepted and understood as a measurement for area, it can be substituted for $\pi/4$ as a representation of area in the flow equation $Q = A \times v$.

In the end, it may be that no single geometry will be adequate for all types of measurement. In looking at flowtime and decimal time, it turned out that decimal time is already used for some sporting events when the situation calls for smaller units of measurement. For example, in the National Basketball Association the last few minutes of the game measure seconds in tenths of a second. In Olympic skiing, some events are measured in hundredths of a second. Conventional time is still in use in most places, however, and this is unlikely to change soon. Likewise, it is hard to argue with Euclidean geometry as a tool for measuring squares and rectangles.

Non-Euclidean geometries have proved to be useful for measuring distances on the surface of a sphere. It may be, then, that Circular Geometry could prove useful for measuring the areas of circles and other round areas. For example, in Figure 7-2 the area of the large circle is 4 round inches instead of π inches. It is all a question of what is the fundamental unit of measurement: the square inch or the round inch. If the square inch is used, then round areas become irrational in area and require the use of π. If the round inch is used, then square areas will also be irrational and will require the use of an irrational value to measure them.

Whether it will work to shift geometries in this way remains to be seen. However, it has worked to use other non-Euclidean geometries when the object being measured is a sphere rather than a plane. So it may also work to measure circles and round pipes using Circular Geometry rather than traditional Euclidean plane geometry.

When I first proposed Circular Geometry many years ago as a solution for measuring the circular area in pipes, an industry colleague commented that "There are so many other uncertainties in flow measurement that the difference in calculated flowrate between traditional geometry and Circular Geometry is negligible." While this is a sentiment worth examining, there is so much value placed on accuracy and precision in flow measurement that I believe that this option is worth exploring.

Units of Area and Volume

Area

1 square inch = 0.45 square centimeters

1 square inch = 645.16 square millimeters

1 square foot = 0.09 meters

1 square yard = 9 square feet

1 square yard – 0.84 square meters

1 square mile = 2.59 square kilometers

1 acre = 4,046.86 square meters

Volume

1 cubic foot = 0.028 cubic meters

1 cubic yard = 27 cubic feet

1 cubic kilometer = 0.24 cubic mile

1 cubic mile = 4.16818 cubic kilometers

1 quart = 0.946 liters

1 liter = 1.057 quarts

1 cup (U.S. dry) = 0.43 pint (U.S. dry)

1 cup (U.S. liquid) = 0.5 pints

1 pint (U.S. dry) = 0.5 quarts

1 cup = 0.5 pint (U.S. liquid)

1 pint = 0.5 quart (U.S. liquid)

1 gallon = 8 pints (U.S. liquid)

Morley's Point:
Area

We still use mathematical concepts that were conceived more than 1,000 years ago. Concepts such as the ratio of the diameter of a circle to the circumference (pi) are still the same. The "irrational" world then was emerging from the "analog" world. Why use sticks to measure a circle?

In physics, surface contains the information. It is not a line nor is it a volume. A black hole has information, but it is inaccessible according to modern physics. Recent suggestions indicate that our universe is merely a surface hologram of our existence. Jesse proposes a new approach to the measurement of surface. Instead of trying to construct every surface dimensionality with a linear approach, he suggests a circular approach. When reading this, at first I thought he was nuts – he's not.

I wanted to make sure, so I contacted a fresh, new, young brain – one enclosed in my friend J. Mark Inman's head. He wrote, "I think it is an interesting conception of pi. It somehow turns two dimensions into one dimension. And while it may not have any practical application (outside of changing convention to eliminate the pi symbol and replacing it with a unit symbol), I think the concept is quite thrilling from the nerd/geek viewpoint, which I tend to share." This seems to be a good idea, but as with most ideas, social magnetism will change our path back to the old. To construct a circle with straight lines is essentially irrational, and the numbers that do it are called irrational. Wire nomenclature uses descriptives, such as 12 gauge, 16 gauge, etc. We should accept that as a measurement system as it is already in use.

Let's talk about line widths and surfaces. I think line width is normally insignificant (zero) compared to an area. We have to deal with the real world, not the mathematical world of 1,000 years ago. A thousand years ago you couldn't get across a room, according to logic. A wide line of demarcation is large but not significant for 100 acres. The DMZ is 2.5 miles wide, but is insignificant for measuring the differences between countries. I know, many fight over these "soft" ratios, but so what. Irrational numbers, such as e, pi, and the square root of minus one, link together to make a rational sum – e to the power i/pi is minus one. As an undergraduate, this shook my world. I felt that this was impossible. We don't need irrational numbers in the real world.

In the physics book *How the Hippies Saved Physics: Science, Counterculture, and the Quantum Revival* by David Kaiser, the author describes the change from the constraining linear physics approach to one of understanding instead of detailing. It suggests that we should not "shut up and calculate" our world. It really does not matter whether the basis of the metric system is a finite dimension associated with a piece of material in Paris, but what does matter is that you understand what it means in the real world. We cannot make the world what we wish it were; we have to deal with the world the way it is. As Jesse says, "You can't fit a square page into a round hole."

This is a very thought-provoking subject – surprisingly so. Congratulations, Jesse.

"It was a great step in science when men became convinced that, in order to understand the nature of things, they must begin by asking, not whether a thing is good or bad, noxious or beneficial, but of what kind it is? And how much is there of it? Quality and Quantity were then first recognized as the primary features to be observed in scientific inquiry." – James Clerk Maxwell

Chapter Eight

Theory of Sensing and Measuring – A Fluid Tale

So far in this book we have looked at different types of sensors and measurement. This chapter builds on the preceding chapters by presenting a theory of sensing and measuring. We will attempt to derive what is common to the examples in those chapters and develop a theory that explains what is essential to something's being a sensor and what is essential to something's being a measurement device.

James Clerk Maxwell

As you will recall, Chapter Four discussed the many types of instruments used to measure flow in closed pipes and open channels. These instruments contain both sensors and transmitters. Chapter Two described the various types of sensors used in the measurement of temperature. These include thermocouples, RTDs, thermistors and optical sensors. Chapter Three described pressure transmitters, which contain both transmitters and pressure sensors, including capacitive, piezoresistive, etc. All of these sensors sense a particular physical state or condition and respond in a predictable way, depending on the quantity or quality of the sensed element.

Chapter Five began by describing how human beings have measured years, weeks and months over the past several thousand years. It discussed the development of today's Gregorian calendar, as well as several other types of calendars, including the Islamic calendar. The chapter then described the historical development of the means to measure time, principally sundials, hourglasses, mechanical clocks, electric and electronic clocks and atomic clocks. It then discussed decimal time as opposed to traditional time, and proposed a new form of decimal time called *flowtime.*

Chapter Six began with a discussion of how our common length measurement units such as the foot, yard, and meter developed. Some of this chapter was devoted to a consideration of Zeno's Paradox, which has important implications for the number line and for the concept of a point. The main argument developed in this chapter was that saying that lines have no width and points have no area leads to contradictions. This chapter laid the groundwork for a different conception in which lines have width and points have area. It also highlighted the need to specify a unit of measurement when making a measurement, in order to avoid a seemingly infinite regress.

The topic of Chapter Seven was the measurement of area, especially of squares and circles. It looked at the fundamental incompatibility between measuring square and rectangular areas and measuring circular areas. As an alternative to using the irrational value of π to measure circular areas with a square unit, this chapter proposed the use of the round inch as a fundamental unit of measurement for measuring circular areas. This is a more theoretical treatment of the old saying that you can't put a square peg into a round hole. The chapter included some axioms of Circular Geometry.

Two Fundamentally Different Concepts

And now we have arrived at our current chapter, where we present a theory that explains the essential difference between sensing and measuring.

It is important to understand that even though they are somewhat related, sensing and measuring are two fundamentally different concepts. A sensor has the ability to detect some quality or property or some object and create a representation of its quantity or quality. This representation varies with the type of sensor and is generally in the form of a physical amount or physical parameter. This representation can then be interpreted as a reading of the quantity or quality of the property being sensed. For example, the mercury in a liquid-in-glass thermometer expands as the temperature increases, rising higher in the tube, and in this way the mercury senses temperature. The resistance of a resistance temperature detector (RTD) changes with changes in temperature, and in this way the RTD senses temperature. A photosensor motion detector sends out a beam of light to a sensor on the other side of a passage or space. When this light beam is interrupted as someone walks through it, the sensor sends out a signal to ring a chime or bell, or increment a counter. Here the light sensor functions like an on-off switch, depending on whether it is sensing light or not.

Measuring is very different from sensing. A sensor has an output that varies according to the presence of some objective quality or quantity of a property. On the other hand, measuring involves determining the size, length or quantity of something in terms of a standard unit. The two main ways to measure something are to compare it to a measuring device that is marked with a standard unit of measurement or to use an instrument that is designed to determine the size, length or quantity of something in terms of a standard unit. Anything that performs the function of measuring is called a *measuring device.*

As we discussed, a mercury thermometer is a sensor because the height of the mercury in a glass or plastic tube varies with the temperature of the mercury. But a mercury thermometer is also a measuring device. The degree units on the tube, whether Fahrenheit or Celsius, register the temperature. Most thermometers have a zero point, and then go into negative or positive territory. Each degree unit corresponds to some portion of an inch of mercury. A mercury thermometer measures temperature because the height of the mercury column can be compared to the position of the standard degree units on the tube. The temperature can be read off the tube.

For a thermometer to be accurate, it has to be calibrated so that the degree units correspond to the actual ambient temperature. For example, 32°F is the freezing point of water, so an accurate thermometer registers 32°F at the freezing point. Likewise, it should read 90°F when the air surrounding the thermometer is at 90°F. So a thermometer is both a sensor and a measuring device.

Another example of a measuring device is a yardstick. A yardstick has 36 inches marked off by thin vertical lines, and each inch is typically subdivided into sixteenths of an inch. Yet a yardstick does not sense the length of (for example) a table; it merely indicates the table's length when it is laid on the table. If it were a sensor, it would have a quality or property that varied with the length, like the mercury in a thermometer. Instead, there is no interaction between the length of a table and the units on a yardstick. The length can be read off the yardstick by comparing the position of the vertical lines to the portion of the table being measured.

In this context, it is interesting to look at clocks. As we asked in Chapter 5, do clocks sense time, or only measure time? It might seem as if clocks sense the quality of duration and respond accordingly, acting as sensors of time. Yet the movement of clocks is independent of duration, and is driven by a pendulum, a battery or by electricity. The hands on a clock or the numbers on a digital clock move in response to their power source, not in response to the quality of duration. Clocks that gain time do not do so because they sense a reduction in duration; they are just incorrectly calibrated or driven to move too quickly in proportion to the duration of time. Hence clocks are like yardsticks of time, and they do not sense the time.

Theory of Sensors: What Is the Essence of a Sensor?

There are many types of sensors and it is not clear that it is even possible to develop a theory that accounts for all of them. Flow sensors typically sense an aspect of fluid flow: some quality or property of the sensor varies according to flow velocity. This quality or property is detected and amplified by a transducer, converter or transmitter. Generally speaking, this information is taken into account along with other variables to produce a reading of flowrate.

Temperature presents a different but parallel situation. One of the most common types of temperature sensors is the liquid-in-glass thermometer. As we have seen, this thermometer takes advantage of the fact that mercury expands as the temperature increases, and contracts as it decreases. By putting mercury in a vertical tube with degree markings, it is possible to determine temperature from the height of the mercury, assuming the thermometer has been properly calibrated. The more sophisticated resistance temperature detectors, or RTDs, make use of the fact that resistance to the flow of electricity in a wire changes with temperature.

On first examination, a sensor with a transducer, converter or transmitter appears to have the following elements:

1. It is composed of some kind of physical material.
2. This material is sensitive to changes in a physical quality or property.
3. The sensing material responds to changes in the qualities of a physical property in a predictable way.
4. A converter, transducer, or transmitter converts these changes into a reading of flow, temperature, pressure or whatever the sensed variable is.

Not every type of material can be a sensor. In order for something to be a sensor, it must respond in a predictable way to the presence of the object or property it is sensing and it must have a specific kind of response to what it is sensing. For this reason, a bar of steel is not a light sensor because it does not respond or react in any way to the presence of light. A stick of wood is not a flow sensor because it simply deflects the flow; the stick is not modified according to the amount of flow.

At the very least, the relationship between a sensor and the sensed object or property has to be a predictable relationship. A mercury thermometer whose readings vary wildly and inconsistently with the temperature could hardly be said to be sensing it, even if it is responding to it. An outside thermometer that reads 60 degrees when it is 20 degrees outside and 90 degrees when it is 35 degrees is not sensing the temperature, but instead responding in a seemingly arbitrary way to it. The idea of sensing contains the idea of truth, so that a sensor must provide an objective value that is within certain bounds of correctness. A thermometer that is a few degrees off can still be said to sense the temperature even if it is not completely accurate.

So in addition to a predictable relationship with the sensed object or property, a sensor must have an implied element of truth or accuracy. This element means that the sensor is acting according to a rule in the presence of the sensed object or property. This rule may not always be known or discovered, but it must exist. It is the rule that formulates the predictable relationship between the sensor and the sensed object or property. In the case of a mercury thermometer, when the mercury is at a certain height, it reads 40 degrees, and when it is at a different height it reads a different temperature, such as 70 degrees. In this example, the rule is that the height the mercury rises to in response to temperature depends on the thermal expansion properties of mercury.

This understanding enables us to formulate a fifth principle that relates to sensors:

5. When a sensor senses the presence of an object or property, it is acting according to a rule that formulates the relationship between the sensor and the sensed object or property. This rule may or may not have been discovered or explicitly formulated.

The Evolving World of Sensors: Mechanical, Electronic, and Biological

There are many types of sensors. We examine three types in this section:

- Mechanical
- Electronic
- Biological

Mechanical

Perhaps the most basic types of sensors are mechanical sensors. There are many types. Before the advent of electricity and electronics, most sensors were mechanical. Mercury in a glass thermometer is a mechanical sensor. A Bourdon tube is a mechanical device for sensing pressure. It consists of a coiled or semicircular tube that is attached via a linkage to a gage that indicates how much the tube is straightened by the pressure inside it. In this case, the position of the free end of the tube responds in a predictable way to the amount of pressure inside the tube. One type of liquid level sensor contains a device that moves along a stem according to the level of the liquid. When the device reaches a certain height, a contact switch closes and sets off an alarm or indicates a value. This switch may be a magnetic switch. Though this type of sensor is an exception, one of the disadvantages of mechanical sensors is that for the most part they have to be read manually.

In flow, pressure, level and temperature, mechanical sensors still exist. A variable area flowmeter is a mechanical sensor that measures flowrate based on the level of a pointer with gradations that typically measure gallons per minute. While most variable area flowmeters must be read manually, some have been developed with an output signal. Mechanical Bourdon tube pressure gages with a dial are still used to measure pressure. Floats are used to measure level, though some may generate an output signal. The level in a tank is measured by the position of a float mounted on a vertical shaft. In temperature, liquid-in-glass thermometers are still very popular ways to measure temperature, especially in non-industrial environments.

Electronic

One of the major advances in sensing in the past 50 years has been the growth in electronic sensors. Electricity has been around for several centuries. Many famous scientists are associated with the development of electricity, including William Gilbert, who is credited with introducing the term "electricity" into the language. In 1660, Otto von Guericke invented a generator that produced static electricity. In 1729, Stephen Gray discovered the conduction of electricity. Benjamin Franklin discovered that lightning was electricity in his famous kite-flying experiment. Thomas Edison made a great step forward in harnessing electricity when he invented the electric light bulb in 1879.

While electricity and electronics are not the same, there is a close relationship between the two. Electricity involves the flow of electrons, and the term "electronic" is derived from the term "electron." Electronic systems rely on electricity, but they also include more complex circuits that combine to create or manage information. A light bulb is an electric device, as is a toaster, while a computer, a cellphone and a stereo system are electronic devices. Electricity often involves high power, while electronic devices are more likely to involve low power.

The advent of electronics and computers has revolutionized the world of instrumentation. Now instead of flowmeters and other devices being mounted in isolation, they can be part of a network of instruments. Many instruments have a transmitter that amplifies the signal from the sensor and outputs it to a controller or recorder. Some instruments output their signals to a programmable logic controller, which can turn instruments on or off, or modulate their output by the use of valves or other input/output devices. Transmitters also have the capability of storing values in memory and of trying to maintain a desired setpoint, depending on the software they contain.

The relationship between electronic transmitters and sensors is important to understand. A transmitter is so-called because it normally outputs or transmits a signal. This signal contains the value of the process variable that is being sensed. In a process environment, the output signal often takes the form of 0–5 millivolts (mV), 4–20 milliamps (mA) or a digital signal. This digital signal could be HART, FOUNDATION Fieldbus, Profibus, or any of a number of other communication protocols. A HART signal is actually superimposed on a 4–20 mA signal, while the digital output signals stand on their own.

Another important component of most electronic transmitters is a transducer or converter. The output signal from a sensor is often very weak, and needs to be amplified if it is to be transmitted. This job of amplification is done by an amplifier, converter or transducer. These devices take the signal from the sensor as input and amplify or convert it to a form in which it can be output more easily. The relationship between a sensor and the physical property that it represents remains unchanged. However, the value of the output signal from the sensor is amplified so that it can be transmitted to another instrument or to a field device.

Biological

Human beings have five main sense organs: eyes, ears, nose, tongue and skin. These sense organs are sensors that accept input from the outside world and convert it in a predictable way to a form the human brain can process and the human mind can understand. All human sense organs are extremely complex, but the eyes and ears are especially complex. Even so, in any given instance the same input is converted reliably into the same mental representation, assuming the sense organ is working correctly. So when someone sees a red object of a particular shade, he or she will have the same or a very similar visual experience as when that shade of red is seen on other occasions. Likewise, the same auditory stimulus will result in the same auditory experience, even though the experience may vary according to direction, pitch and other auditory qualities. A clashing cymbal or a cardinal's cheerful song sound similar each time, although the specific auditory experience varies with the direction and intensity of the sound.

How the mental experience of a sight or a sound or a touch becomes part of not just the brain but of the mind is part of the classic mind-body problem formulated by Descartes. What it makes sense to say about the mind-body relationship in this context is that the brain has extensive processing powers that are enabled by neurotransmitters. These sensory experiences are processed by the brain in a way that is somewhat similar to the way in which a transmitter processes a physical sensory signal. Of course, a transmitter is electronic, while the brain is biological, organic, neurological – even electrical – and is linked to consciousness.

Descartes set up the mind-body problem in a misleading way. The problem as it is often conceived of is to explain how an immaterial and nonphysical mind can interact with a material and physical body. Descartes chose the pineal gland in the brain as the point of interaction because, unlike the five main senses there is only one pineal gland in the brain, while the five senses each use two or more processing centers. While subsequent science has not confirmed Descartes' view of the role of the pineal gland, this does not undercut the great service he did in setting up the problem.

An Alternate Solution

Descartes defined the body as a material object and the mind as an immaterial object. By defining mind and body as complete opposites, Descartes made the mind-body problem impossible to solve within this framework. Instead, if we treat the mind as having some physical properties and the brain and the body as having some mental properties, then their interaction becomes easier to account for. When I have a thought that I will pick up a glass, the physical aspect of my thought initiates a series of neural and bodily responses that results in the physical act of my picking up the glass. The physical aspect of my thought is the brain activity that is associated with my having the thought. Because this neural activity occurs in the brain it triggers a series of neural events that causes my body to engage in the activity I am thinking of. The series of neural and bodily responses is one that is learned over time and repeated many times as the occasion arises. In this way, my thought can be said to initiate a physical action.

The above example shows one way that the mind can influence the body. But mind-body interaction is a two-way street. When someone gets a cut on their hand, sensory impulses are sent to the brain. A series of neural impulses that are associated with a feeling of pain are processed in the brain. The result is a feeling of pain. The sensation of pain has both a physical and a mental aspect, and in this way the body can be said to interact with the mind and to cause the feeling of pain.

Can Robots Feel Pain?

Much has been made in science fiction of the idea that robots are intelligent creatures and that some day they may become so smart that they will attain consciousness, outthink their creators, and even come to rule them. This seems unlikely. Robots are mechanical and electronic creations and lack the organic and neural correlates required for consciousness. While it is not easy to explain this link, there appears to be a necessary link between biological and neural structure and consciousness. Unless robots somehow become biological organisms that are born and die, they will never attain consciousness, even if they can be programmed to mimic the expression of pain. Likewise, other electronic structures such as flowmeters do not feel pain when they are damaged or angry when they are turned off. What sometimes makes excellent science fiction does not always make science fact.

Theory of Measurement: What Is the Essence of Measuring?

As we have discussed, measuring is quite different from sensing. Sensing requires the presence of a sensor that responds in a predictable way by creating a representation of some quality or property according to an implicit or explicit rule. By contrast, measuring is a simpler operation. Measuring requires a unit of measurement, and this unit of measurement is used to determine a quantity that formulates how much of that unit something has.

Probably the biggest difference between sensing and measuring is that there is some type of interaction between a sensor and what is sensed. With measuring, this type of interaction does not occur. Instead, a measuring device determines how much size, length or other quantity of a unit of measurement something has.

One of the clearest ways to understand the nature of measuring is to consider a yardstick. The unit of measurement for a yardstick is an inch, although it is also divided into three feet and into smaller increments of an inch. As we have seen, a yardstick is designed to measure length, an abstract property defined as the distance between two points, one at either end of the object being measured. To measure the length of an object using a yardstick, the yardstick is held next to the object. The length of the object can be determined by comparing the lines on the yardstick to the object at the points that lie at either end of the measured length. There is no interaction between the object or its length and the yardstick; it is a simple comparison.

The choice of a unit of measurement is somewhat arbitrary, although as we have emphasized, it has to be appropriate to the property or quality being measured. In Chapter Six, we saw that the length of a yard was determined in the 12th century to be the distance between King Henry's nose and the tip of his out-stretched thumb. Also during this time, the idea arose of a foot being 1/3 of a yard. The definition of a foot as consisting of 12 inches, however, goes back to Roman times.

Once a unit of measurement is defined, other units can be defined in terms of it. So 12 inches make a foot, 3 feet make a yard, and 5,280 feet make a mile. The unit of measurement used also needs to be appropriate to the quantity of the desired measurement. Because of the magnitude involved, the distance from the earth to the sun is usually given in miles rather than feet or inches. At the other end of the spectrum, it is impractical to measure the height of a desk in miles, partly because there is no "milestick." (However, such a distance could be calculated. For example, the height of a 30-inch desk is about 1/2,000 of a mile.)

Different countries and traditions use different units of measurement. In the United States, the units of inch, foot, yard and mile still predominate. In much of the rest of the world, especially in Europe, the units used are the meter and kilometer. A millimeter is 1/1000 of a meter, and is equivalent to 0.03937 inches. One inch is about 25.4 millimeters.

In many cases there is no exact conversion between the American system and the metric system, only approximations. Sometimes this may not matter so much, but other times it does. For example, a car made according to metric measurements cannot easily be worked on by someone using American-defined wrenches; the wrenches won't quite fit. In market research, however, these differences are often ignored, depending on the intended audience. When reporting on the diameter of an instrument such as a flowmeter, for example, one unit or the other is taken as the standard, and a conversion is assumed. So a flowmeter that is 25 mm in diameter is treated as a one-inch meter, and the same logic is applied to the other sizes.

In Chapter Six, we looked at various attempts to define the length of a standard yard. This included an instance where the standard yard bars were destroyed by fire. Eventually, there was a similar attempt to define the standard meter. In 1983, the idea of a physical bar was abandoned in favor of a definition referencing the distance that light travels in a vacuum in a very small amount of time. The standard yard has now come to be defined in terms of the length of a standard meter, making the meter the primary standard and the yard a secondary or derived standard.

A similar logic applies to other units of measurement, such as ounces, pounds, square inches, acres, seconds, barrels, gallons, grams and teaspoons. In each case there is a fundamental unit of measurement that there is one of, and larger units are defined in terms of that fundamental unit. Defining the exact quantity of that unit of measurement can be challenging, as we saw in the case of the yard and the meter.

Weight is another example of the difference between the American system and the metric system. The American system uses ounces and pounds, while the metric (SI) system uses grams and kilograms. One ounce equals 28 grams. Most of the world, however, has standardized on the second as the fundamental unit of time, with 60 seconds in a minute, 60 minutes in an hour and 24 hours in a day. However, several attempts have been made to adopt a decimal system for time, and some of these were examined in Chapter Five.

Simple Comparison Devices: Yardsticks and Dipsticks

Much of the measurement that occurs in ordinary life is done by simple comparison devices, such as the yardstick. The lines on a ruler or yardstick are compared to a desired length, and the length is read off the ruler or yardstick. Time is read off the hands of a clock, which is usually marked in hours and minutes. The amount of coffee in a carafe is read off the scale on the side, usually indicated in cups. An oil dipstick registers the amount of oil in a car's engine by a comparison of the oil level indicated on the stick with the line on the stick indicating Full and (most often) one quart low. Measuring spoons and cups are constructed to contain a predetermined amount of liquid or solid, and these are simply filled exactly to the top when they are used in baking and cooking to measure out quantities.

What is common to these simple comparison devices is that they incorporate a standard unit, often in the form of a scale, and they measure size, length or quantity by a simple comparison of the scale or marked units with the object or fluid being measured. These devices are read manually, and no calculation is required. While these simple comparison devices are indispensable in ordinary life, they do not always satisfy the more rigorous requirements of scientific or industrial contexts.

More Complex Measuring Devices: Meters

A short definition of a meter is that it is a measuring device. A longer definition of a meter is that it is an instrument designed to measure the size, length or quantity of something. An instrument is a device that is designed to perform one or more specific tasks that may involve detecting, signaling, communicating, observing, recording or measuring some quantity or phenomenon; controlling or manipulating another device; or performing a similar function. Examples of instruments include tachometers, flowmeters, pressure transmitters, barometers, telescopes, colorimeters, fluoroscopes, gravimeters, manometers, odometers, oscilloscopes, recorders, psychrometers, spectrographs and wavemeters. While not all instruments (such as valves) are measuring instruments, many incorporate measurement into their functions.

Flowmeters are a type of meter, even though they may not have a scale displayed. The scale or value incorporating a unit of measurement is generally displayed in the flowmeter's transmitter, but not all flowmeters have transmitters with a display. Of course, mechanical flowmeters such as variable area flowmeters do contain a scale, generally indicating gallons per minute (gpm). The scale indicates the flowrate of the fluid flowing through a pipe, and flowrate is read off the height of the moving float. More modern flowmeters use more complex technologies to determine flowrate, and show the value in an electronic display. These flowmeters incorporate a sensor of some type to sense the fluid flow, then compute the flowrate, often using other variables such as inner pipe diameter. Different types of flow sensors include ultrasonic, magnetic, thermal, differential pressure and others. Some flowmeters take volumetric flow, temperature and pressure into account, and measure mass flow as well as volumetric flow. So flowmeters are correctly considered meters, since they take a sensed value and convert it into a value that incorporates a unit of measurement. The different types of sensors used by flowmeters are described in Chapter Four.

What Is the Essence of a Measuring Device?

Most measuring devices are either simple comparison devices or they are types of meters. What are the common features of these measuring devices? Like sensors, they have a common set of characteristics:

1. A measuring device makes use of a standard unit of measurement representing some quantity.
2. Additional units of measurement are defined using the standard unit of measurement and are either marked on the measuring device like a scale or are calculated based on the unit of measurement.
3. The measuring device measures the quantity of units the object or fluid has when the units scale is compared to the measured object or fluid, or the units are calculated from relevant properties of the measured object or fluid.

Different Types of Meters

There are many different types of meters that measure different quantities. Many of these are used in scientific and industrial circles, though some have uses in everyday life.

The following is a list of some of the more common meters and what they measure:

- Accelerometer: Measures the acceleration of aircraft or rockets; also now used in smart phones.
- Altimeter: Measures the height above ground.
- Barometer: Measures atmospheric pressure.
- Calorimeter: Measures quantity of heat in a substance or object.
- Densitometer: Measures optical or photographic density.
- Hygrometer: Measures the humidity of the atmosphere.
- Magnetometer: Measures the intensity of magnetic fields.
- Micrometer: Measures very small distances.
- Odometer: Measures distance traveled.
- Oscilloscope: Measures electrical fluctuations.
- Photometer: Measures light intensity.
- Piezometer: Measures pressure or compressibility, especially water pressure in a land mass.
- Solarimeter: Measures solar radiation.
- Turbidimeter: Measures turbidity of liquids.
- Voltmeter: Measures electric potential.

Definition of Instrument: A mechanical or electronic measuring device used to sense or determine the flow, pressure, temperature, level, speed, position, or similar quality of an object or physical device.

Conclusion

The world of sensors and measurement is essential to both engineering and daily human life. This book has traced the fascinating development of some of our most important units of measurement, including length, area, and time. It has shown the arbitrary nature of some of these units, yet this does not make them any less useful.

The first half of the book discussed some fundamental methods of sensing, including temperature, pressure, and flow. Like the units of measurement, these methods of sensing are indispensable aspects of engineering and daily life. Yet this is a fluid tale. Sensors are becoming more and more prevalent in our society, and new methods of sensing are still being invented. Likewise, new types of meters are finding their way onto the market. Hopefully the theory and definitions in this chapter will accommodate these new developments. Whether they prove adequate, or need to be revised in light of further developments, the world of sensors and meters will always be a fascinating and challenging one.

Morley's Point:
Theory of Sensing and Measuring

Chapter 8 is mostly about philosophy and the difference between sensing and measuring. Everything in the universe is a sensor. Galaxies respond to other galaxies by sensing gravity and quantum pressures.

Most of what is sensed is difficult to measure. As an example, how do you feel? Are you healthy? We use temperature, blood pressure, and heart rate to establish the wellness of a patient – not precise, but good enough.

In engineering we try to get sensing elements that are repeatable and have a wide deviation as a result of changes to the stimulus. Because we now have computers, the results need not be linear. They do, however, need to be consistent. Sensing and measurement is a fundamental philosophy that needs to be embedded strongly into the reader's brain. Although everything is a sensor, not all sensing elements are useful in the engineering world.

On a personal note, I have no sensing of pain in parts of my body. I cannot tell if I am injured because I do not have the warning of pain. What I have to do is make sure my blood pressure is high enough. If it goes too low, it means that I have pain even though I'm not aware of it. Normally your brain is the measuring device, but my measuring device is flawed so I use blood pressure as a gauge. It's not very good because it's not quick.

Philosophically, sensing for humans use five or six senses. Some people include the ones we know about and consider the brain to be another one of the sensing elements. I, however, would argue that the brain is a measuring element, not a sensing element.

The sensing capability of an animal (which includes us) is impressive. We can sense and measure acceleration, attitude, danger (fight or flight), space occupancy and balance. The list is almost endless. Animals have neurons for all senses that we can dream about, like hearing and pressure, while some even have magnetic sensitivity (some humans are sensitive to electric fields as well). And the beat goes on.

The main argument is that anything can be a sensor. The useful ones are repeatable and have a wide range of sensitivity. Measurement is the ability to take that sensor and make some rational sense of what it is sensing so we can put it in a useful application. Sometimes we are overwhelmed by the illusion of accuracy, so caution is advised. Most of us have a temperature gauge as an accessory for our car. The number displayed is based on a sensor, and the small low-cost element tells us whether we should be comfortable. It's amazing how we convert the numbers in our brain from another sensor and measuring device to obtain the degree of comfort. If the car says 72°F, I should be comfortable.

In summary, anything can be a sensor, but the useful ones have a wide range of repeatable performance. (For instance, we can measure the wobble of a star to detect captive planets.) Remember, sensors are first in line in the measurement process.

Morley's Final Point:
Futures in Measurement

The measurements of the future will have to take into consideration all the physical characteristics of the material being measured. They can be part of making pills, automobiles, water/wastewater or fuels. You understand most of the history and the near-term future of measurement from the major chapters in this book. We now lightly touch on some possibilities of future measurement techniques.

There is an obvious trend to reduce the size of sensors and increase the sensitivity of measurements, and significant work is being done to measure gas concentration and liquid density. We now measure the things that earlier were abstract such as density, measurement, and composition of a liquid.

We particularly need to know what the composition is of the material being processed. We have to know its mass, weight, volume and unwanted contamination. Single measurements won't cut the mustard. The ability to infer composition by multiple simple measurements in a single elemental sensor will be offered. Sensors will be able to log groups automatically and be able to send these results up to the cloud of complexity that will replace most physical arrays.

This means that each sensor element, whether complex or simple, will have an interface to a standardized cloud array that will represent the traffic manager of information for the system. One of the more interesting developments is the carbon tube nano sensor being used for pressure.

Stacked carbon tubes — with one end open — get significant results for very small pressures. Who knows? You may be able to build a sensing element system with a desktop computer. Printers used to be the size of my barn. Now, they are the size of my suitcase. The actual size of processes and their sensors will diminish rapidly. 3-D manufacturing is not just for solids, but can also be done with liquids. Think of a USB connection to a box that makes perfect martinis.

As we go to nano sensing and semiconductor analysis, we do not need linear sensing elements. As long as the setting is predictable and has a sweet spot of high resolution, small microcomputers can take care of the data manipulation to emulate a linear output. And as long as it's repeatable, computers can make it happen.

Sensors will be smaller, lower cost, computer compatible and capable of merging their outputs with other similar sensors to get results. The work of assessing an analysis will be about 80% software and 20% hardware. This means small companies will be designing apps for processing fluids and will grow using these techniques of composition measurement, becoming a great source of innovation. History tells us that we cannot discount the innovation inherent in most modern cultures.

Some of the Physics

Most of our discussion about sensing in this book is directly coupled to measurements. We have discussed repeatedly that sensing and measurement are different. Sensing is the distortion of an element by the environment. Measurement is the numbers for that distortion. The future physics for automation is not in the nano region — particularly for the short run.

Quantum mechanics is a statistical analysis and has no firm discrete measurement. This will hurt the heads of most engineers, but our ability to think about macro physics will help us in the short run. For example, electromagnetic coupling, capacitance, etc. will be important in the next five to 10 years. The physics will allow smaller, lower cost, higher accuracy measurement of almost any of the required dimensions for fluid.

In addition, we will measure the major constituencies, impurities, energy content and mass in most processes. Sometimes I think we are going too far; I just want to know what is coming out of the faucet. But who am I to suggest that? The macro physical characteristics of materials will be directly measured by new semiconductor devices and the ubiquitous computers are everywhere. To repeat, a measurement that is repeatable, independent of linearity, is useful to us in this computerized world. Feedback around these control loops is a key element to be considered. It is possible to get hysteresis and other noise in the measurement. We have to be aware of the measurement time and the bandwidth of the process.

There is a subject matter which we have not approached much in this tome — the system itself. The system consists of operators, power, environment, etc. Operator alarms, for example, are a key to reliable performance. At Three Mile Island, for example, there was only "one" event, but all alarms went off without anyone paying attention to the sequence. The ability to systemize the connection between operator and output is a key element in these considerations. Operator training required risk analysis cost of product delivered — not cost of the components in the system — a key element. I'm reminded of town meetings in our little town of Mason — about 1,200 people. I went to one town meeting and they spent 15 minutes deciding on a million-dollar budget for school. And they spent another 15 minutes deciding whether to put new tires on the police cruiser. It's funny how we treat all problems equally, independent of their actual importance.

I teach all over the world (distance learning) and make this point: Don't let the accountants make you decide what product to buy. Buy the product that optimizes the system. In the automotive automation market, the manufacturer can make a pickup truck every three minutes. If the system is down for a day, that's a lot of trucks not being made. Think about eliminating the midnight call to you to fix the system tonight.

Well, this is the end of our story. Stay warm and have pleasant dreams. Tomorrow is another day.

Bibliography

Boorstin, Daniel J. *The Discoverers: A History of Man's Search to Know His World and Himself.* New York and Toronto: Random House, Inc., 1983.

Hebra, Alex. *Measure for Measure: The Story of Imperial, Metric, and Other Units.* Baltimore, MD: The John Hopkins University Press, 2003.

Klein, Herbert Arthur. *The Science of Measurement: A Historical Survey.* New York: Dover Publications, Inc., 1974.

Spitzer, David W. *Industrial Flow Measurement.* Research Triangle Park, NC: ISA, 2004.

Upp, E. L. *Fluid Flow Measurement: A Practical Guide to Accurate Flow Measurement.* Houston, TX: Gulf Publishing Company, 1993.

Yoder, Jesse. "Coriolis Effect Mass Flowmeters." *Measurement, Instrumentation, and Sensors Handbook, 2nd Edition: Spatial, Mechanical, Thermal, and Radiation Measurement.* Ed. John G. Webster and Halit Eren. Boca Raton, London, and New York: CRC Press, 2014. Section VI, pp. 60-1–60-6

Yoder, Jesse. *The World Market for Infrared Thermometers and Thermal Imagers.* Wakefield, MA: Flow Research, Inc., 2002.

Yoder, Jesse. *The World Market for Pressure Transmitters,* 4th Edition. Wakefield, MA: Flow Research, Inc., 2014.

Yoder, Jesse. *The Market for Temperature Sensors in the Americas, 2nd Edition.* Wakefield, MA: Flow Research, Inc., 2006.

Yoder, Jesse. *The Market for Temperature Transmitters in the Americas, 2nd Edition.* Wakefield, MA: Flow Research, Inc., 2006.

Yoder, Jesse. "Ultrasonic Flowmeters." *Instrument Engineers' Handbook, 4th Edition: Process Measurement and Analysis, Volume I.* Ed. Béla G. Lipták. Boca Raton, London, and New York: CRC Press, 2003. 2-26, pp. 353–361.

Yoder, Jesse. *Volume X: The World Market for Flowmeters, 5th Edition.* Wakefield, MA: Flow Research, Inc., 2014.

Index

G

N

O

W

Y

Z